HOW I LEARNED TO RELATE TO MY LABORATORY RAT THROUGH HUMANISTIC BEHAVIORISM 2.2

A Laboratory Manual

Richard W. Malott
Websites: www.DickMalott.com, www.Behaviordelia.com

Behaviordelia
8971 W. KL Avenue
Kalamazoo MI 49009
USA

(269) 372-1268

How I learned to Relate to my Laboratory Rat Through Humanistic Behaviorism 2.2
New Edition: 2012

ISBN: 0-914-47400-6

Version 1.0 prepared by

Dr. Richard W. Malott

Dr. Paul Mountjoy
Daniel Fritcher
Cathy Janczarek
Sandy Laham
Ronald McGory
Kathy Rachford
Daniel Reese
Kristi Ritterby
Richard Woolfenden
Ronald Borne

Version 1.1 prepared by

Wendy E. Jaehnig

Version 1.2 prepared by

Amy E. Scrima

Version 1.3 prepared by

Kip DenHartigh

Version 2.0 prepared by
Jason Otto
Moira McGlynn
Tracy Brandenburg

Version 2.1 prepared by
Jennifer Skundrich

Version 2.2 prepared by
Jessica Korneder

Art

Patricia Hartlep
Stuart Hartlep
Terry Boothman

Photographs

Kip DenHartigh

Tyler's Tales

Lori Johnson (December 16, 1992)

This laboratory began, in the Introductory Psychology course and is still used at Western Michigan University. Data sheets are included with each experiment, which students turn in upon completion.

PREFACE

We have had an operant-conditioning rat lab associated with either our Experimental or Introductory Psychology course at Western Michigan University since 1964. Many students report that the rat lab is very exciting and that they find it the best feature of our course. It has been our impression however that this lab has an even more important effect. The experience of directly modifying the behavior of another organism is a primary factor in demonstrating to students the power and validity of the basic principles of behavior.

TABLE OF CONTENTS

INTRODUCTION

How I Became A Scientist and Loved It!

This series of laboratory experiments will give you the opportunity to identify, predict, and influence the behavior of another living organism. You'll come to develop a warm and close relationship with (put name here), your laboratory rat.

There will be questions within the text, and also after the different sections in this manual, including each lab. Be sure that you answer ALL the questions in each assignment before coming to class. **Your lab instructor must check and sign off on these questions before you can start your experiments.** There will also be questions after each experiment. You will not complete these until you have completed the experiment.

YOUR FRIEND or Will I really hit it off with my friend, the (ech!) laboratory rat?

You bet you will, and don't think we're putting you on! It's been our experience that participation in introductory science laboratories is a real downer. However, good fortune smiles upon you, because this lab will be tons of fun.

In order to understand your behavior, as well as your rat's, you'll need a small, but precise technical vocabulary. Once you've learned to wield these razor-sharp concepts with skill, you'll be able to cut through the fat and get to the heart of any behavior issue. "Quick," you say, "teach us the first concept!"

That concept is REINFORCER. A reinforcer is roughly the same as a reward. But let's be a little more exact. A REINFORCER is (circle one)

- (a) an elderly door knob collector
- (b) any stimulus, event or condition whose presentation immediately following a response increases the frequency of that response.
- (c) a person who puts metal rods in poured-concrete foundations.
- (d) a person who keeps a careful eye on you *after* you've gotten a speeding ticket.
- (e) another name for "flying buttress."

Did you answer "b?" You did? Well hot rats! You're well on your way to becoming an ace behavior analyst.

We all find it very reinforcing to have some impact on the world, to influence and deal effectively with the environment. You will find it extremely reinforcing to be able to predict and influence the behavior of Robby the rat, or, as he is affectionately known, No. 327J. You will be using water as the reinforcer for your rat.

EQUIPMENT:

The equipment you will be using should be handled with care. Even though you may be an expert mechanic with your car, you are not expected to repair the apparatus. **If some part of the equipment you are using does not operate properly during an experiment, stop and inform your lab supervisor of the condition.**

The apparatus is fairly simple, but before you begin operating the equipment, your lab supervisor will demonstrate its use.

You'll be working with a Skinner box. There are two parts of the box you will be operating:

- (1) a water dipper that will allow you to deliver water to the rat from a water tray (kind of the same way you get a spoonful of soup from a soup bowl).
- (2) a light switch that turns a light on in the Skinner box.

The inside of the Skinner box has a lever that the rat will push.

The only other equipment necessary for your success is a watch with a second hand or a timer/stopwatch of some kind. You will need to provide this and bring it to every session.

THE EXPERIMENTS or What the heck will I be doing?

All the experiments will be more thoroughly explained later in this manual, but here's a quick overview.

In Experiment 1, you'll train the rat to respond to the sound of the dipper clicking up whenever and wherever he is in the Skinner box. You'll do this by pairing that sound with the reinforcer – water.

In Experiment 2, you'll reinforce the rat's lever-pressing response, by delivering a drop of water every time the rat presses the lever. Then you'll extinguish the response by no longer reinforcing it when it occurs (causing the response frequency to decrease).

In Experiment 3, you will reinforce the lever press when the light in the Skinner box is on, and not when it's off. This is called discrimination training. Eventually, the rat will only press the lever in the presence of the light, and not in its absence.

In Experiment 4, you'll hang a chain through a hole in the top of the Skinner box, where your rat can reach it. Then you'll reinforce a chain-pulling response by turning on the light when the rat pulls the chain. After the light is on, the rat will press the lever, because in the past, lever pressing has been reinforced in the presence of the light (Experiment 4). You will, of course, immediately reinforce the lever press with the dipper click and then turn off the light. Then the process will start all over again (back to the chain-pull). This is called a stimulus-response chain. Pretty cool huh? Hang on, there's more!

In Experiment 5, you will gradually increase the amount of weight the rat has to push by adding washers to the back of the lever. You may have to name your rat Hercules!

In Experiment 6, you will increase the number of lever presses required before giving your rat the reinforcing dipper click. First you will require 1 response, then 2, then 4, then 6 and finally, 8! This is called a fixed-ratio schedule of reinforcement. Nifty isn't it?

In Experiment 7, the last one (whew!), you have to design your own experiment. Don't worry, at this point you'll be so great at this that you'll be able to impress us with a nifty experiment that would make B.F. Skinner proud!

If you want to, and if you have time, you can design and implement another experiment for optional activity points (OAPs), and impress the heck out of us!

We'll be going over the experiments in some detail later in this book, so don't freak out if you don't understand all the experiments right now.

WHISTLE BLOWING: What to do when your rat isn't progressing through an experiment.

You can ask your TA for help any time you are having problems with an experiment. We also have guidelines for "blowing the whistle." For each experiment, there is a whistle-blowing criterion. If your rat has not completed the lab within the specified amount of sessions, blow the whistle. How do you blow the whistle? Simply let your TA know that you have been on the experiment for X number of sessions, and that you are blowing the whistle. Your TA will help you analyze any problems you're having with the experiment, so you can make progress toward completing it. This will prevent you from getting stuck on any particular experiment, so be sure to take advantage of it.

ANIMAL HANDLING AND CARE: or Don't grab any tail in the lab.

Both you and your rat will do better in the course if you hit it off from the start. You must remember that these animals are naive and have not been handled a great deal. Therefore, it is very important that in your first several contacts with the animal, you handle him gently. Spending some time in the beginning of the first few sessions holding and petting your rat is a good way to assure an amiable relationship. Should the animal be handled roughly or dropped, he will be far more difficult to train.

You may have had a pet hamster sometime during your childhood. The laboratory rat is very similar in nature, since both animals have been domesticated for some time. There is no need to be afraid of the laboratory rats, as they are quite small and harmless. You are many times their size and undoubtedly much wiser. The following are some important points to remember in handling your rat:

(1) There are several ways you can safely lift your rat. One is to wrap your hand around him and lift, with your fingers around his belly. (This is done in the same way you would grasp the handle of a bucket of water.)
(2) When you are holding your rat, be sure to support all of his feet. Nobody likes to have his or her feet dangling in midair.
(3) Hold your arm close to your stomach and let the rat move along your arm. He will probably stop at the crook of your arm.
(4) You may then stroke the animal, advisably behind the ears.

In the past, some students have handled these animals aversively. Below is a list of important points to remember:

(1) Do NOT chase the rat around with your hand when you are trying to pick him up from his cage or the Skinner box! Instead, find your rat before you attempt to pick him up. Once you find him, immediately reach in and pick him up gently yet firmly. Do not panic if he moves. If you end up chasing him around with your hand, that will only make it more difficult to pick him up.
(2) Don't pick up the rat by the skin on the back of the neck, as though he were a kitten. Remember, this is a rat.
(3) Make "friends" with your rat as soon as possible. This will be a tremendously important factor influencing his performance in the lab. Handle the rat as gently as possible, taking care not to make loud noises. Treat your rat as if your grade in the course depends on him.
(4) Once you're working with your rats in the Skinner box, it's important to remember that it is very sensitive to noise and movement. The more still and quiet you can be while training, the faster and easier it will be.
(5) Don't put your fingers up to/through the grates of the cages. The rats don't have great eyesight and may mistake your finger for their food.

UNIVERSITY POLICY: The rat stops here.

Just like the Skinner boxes, stools, and tables in the lab, the rats are property of the university and are **NOT** to be taken from the lab *(even if he is really cute and would be going to a nice home)*. In order to provide students with the valuable opportunity to work with lab rats, the department must follow strict guidelines for maintaining the safety and care of these animals. Allowing the rats to become pets afterwards would violate these guidelines and jeopardize the opportunity for future animal labs. Therefore any stolen or missing rats will result in an incomplete in the class for the student and possible further action by the university.

ONE MORE THING

Because the rats are so sensitive to sound and movement, it makes it that much more important to be sure you're <u>on time</u> for lab. When you come in late you inevitably disturb the others that have already started and of course their rats. This can be very frustrating, not to mention if you're late, you have less time to work with your own rat. This can be a problem because the experiments are set up to be completed within a certain number of 50 minute sessions (basically) and so cutting into that time may screw up your training schedule. Anyway, you get the point – don't be late!

QUESTIONS (always answer ALL the questions)

1. What will you be using as the reinforcer for your rat?
   (a) food pellets
   (b) money
   (c) water
   (d) pizza

2. What part of the Skinner box apparatus will you be using to deliver the reinforcer?
   (a) dipper
   (b) lever
   (c) light

3. How many experiments will you be doing?
   (a) five
   (b) seven
   (c) four
   (d) six

4. When you are trying to pick the rat up from the cage you should chase him around with your hand for a little while at first. This will calm the rat.
   (a) true
   (b) false

5. You should always support all of the rat's feet when you are carrying him.
   (a) true
   (b) false

6. You need to bring your own watch or stopwatch with a second hand.
   (a) true
   (b) false

7. I am **NOT** allowed to take my rat home as a pet or set him free at the end of the semester.
   (a) true
   (b) false

8. When I start my training session I should:
   (a) blow my nose and wave to my rat
   (b) talk to my rat, encouraging his progress
   (c) sleep
   (d) stay as still and quiet as possible, paying close attention to my rat's every move

EXPERIMENT I
Movement and Dipper Click Training

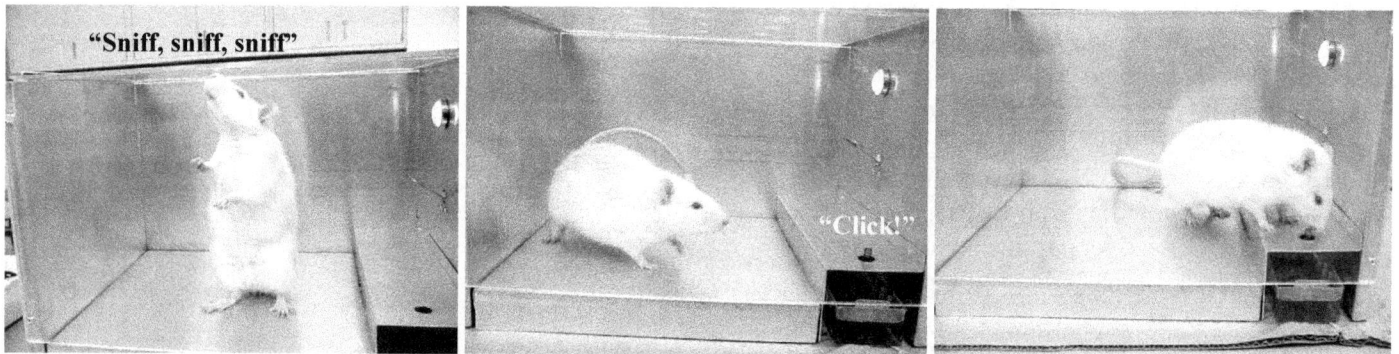

"Sniff, sniff, sniff"

"Click!"

After training, when you raise and click the dipper you will see your rat turn and approach the dipper just like this.

The purpose of movement training is to deliver water after the rat moves about the Skinner box. Dipper training is to train the rat to move to the dipper from any place in the chamber *when you click the dipper,* and only then.

As your rat is on a 23-hour deprivation schedule (he has not had water for 23 hours) it should not take long for him to find the water. Follow these procedures:

(1) Make sure the light in your Skinner box is **ON** when you start this experiment.

(2) Give a drop of water without clicking the dipper. Do this until you get movement from the rat (until the rat goes to the dipper without hearing the click).

(3) Place the dipper in an upright position, but not too hard. You want the dipper to barely make a sound when it comes up so the rat can pair that sound with the presence of the water. But if you raise it too hard and the click is too loud it will be aversive to the rat and you'll have a lot harder time with the pairing. Raise the dipper quietly at first and get louder as the time goes on, but not too loud.

(4) Allow the rat time to find the dipper and lick it for a few seconds. Do not hold the dipper up longer than 5 seconds from the click. If he does not come, click the dipper again—after he moves in some way.

(5) As soon as the rat begins to move away, immediately but gently lower the dipper into the water reservoir (the tray containing the water). Then raise the dipper into position. The dipper will make a slight click when you raise it. Eventually your rat will only approach the dipper after the click. Be careful not to pinch the rat's tongue when the dipper is raised. We want the behavior of approaching the water dipper to be reinforced, not punished!

(6) Once the rat has adapted to the click of the dipper, you may start moving the dipper up and down more quickly.

(7) Eventually, the rat should come immediately to the dipper when the dipper is clicked, no matter what part of the Skinner box he is in.

(8) Most importantly try to deliver clicks immediately after your rat makes some behavior (not when it's sitting, or just swaying near the dipper).

(9) Read through these pointers two more times; this experiment may involve the most crucial learning time for your rat.

The dipper training procedure is also what is called a "value-altering procedure." A value-altering procedure is a procedure in which a neutral stimulus is paired with a reinforcer or aversive condition, and becomes a learned reinforcer or aversive condition.

The two stimuli involved in the dipper training procedure are the "click" of the dipper and the water.

Can you guess which stimulus is the neutral stimulus?
  (a) the "click"
  (b) the water

Of course you correctly said that the "click" was the neutral stimulus. And that means that the water is the reinforcer. During the dipper-training procedure, the click is repeatedly paired with the water.

However, when there is no "click" (when you have NOT raised the dipper), there is also no water.

After repeated pairings of the "click" and the water, the "click" comes to function as a learned reinforcer for the rat. Thus, the sound of the click will reinforce (strengthen) whatever response it follows.

There are some things you should remember when you are dipper training your rat.

(1) Remember, turn the light **ON** and leave it on during training.

(2) Do NOT reinforce the behavior of hovering about the dipper too often, as this is incompatible with the terminal behavior of lever pressing, which you will later establish. If you reinforce this behavior too much, your rat will **never leave the dipper**. You will end up spending the entire lab period watching him stare at the dipper.

(3) Do NOT click the dipper too hard, especially when you are first dipper training. In general, rats (and all animals actually) are very sensitive to sound and movement. **A good rule of thumb is once you sit down to start your training session, get into "position" (right hand on the dipper and left hand ready to write or on the light switch) then keep all sound and movement to a minimum.** If you have to move – do it slowly. Don't talk or tap your fingers or scratch your head for example. This goes for coming into the lab also. Others may have already started and if you push your chair in or start talking to your friends, you may be screwing up their session. Basically, the fewer distractions, the easier it will be on both you and your rat.

(4) **Don't raise the dipper if the rat is standing stationary over the dipper.**

(5) Also, it is okay to inadvertently reinforce the rat's behavior of moving away from the dipper because it is some kind of movement, but do not just reinforce that particular behavior exclusively, at all times. You should reinforce all movement rather than one particular type of movement. If you see your rat doing a specific behavior over and over, don't deliver water for it, wait and click the dipper after he's done a little something different.

(6) When you start to see your rat approaching the dipper more frequently immediately after you click it, you should start recording data.

(7) **Raise the dipper for only 3 or 4 seconds. Do not hold it up longer than 5 seconds.** This is important for two reasons. First, it is more than enough time for the rat to drink the available drop of water. If you leave the dipper up longer, a great deal of time will be wasted because the rat usually continues to lick the dipper until it is lowered. On the other hand, if you lower the dipper earlier than 3 seconds he may not have time to consume the water, and his behavior will extinguish (stop occurring).

Second, if the rat does not approach the dipper immediately after he emits the response, other behaviors that are incompatible with lever pressing may be conditioned. For example, if the rat presses the lever and you immediately raise the dipper, then he goes to the opposite end of the Skinner box before approaching the water, that response will be reinforced, not the lever press.

WHISTLE BLOWING: If you have not completed this lab after two lab sessions, blow the whistle! **In other words, let your TA know so he or she can help you progress through the experiment.**

QUESTIONS FOR EXPERIMENT I

(Remember, you must answer these and have your lab
instructor look at them and initial the space below
BEFORE you start the experiment.)

Instructor's initials: _____

(1) What is the purpose of dipper training?
    (a) to adapt the rat to the Skinner box.
    (b) to train the rat to move to the dipper from any
    place in the chamber when you click the dipper,
        and only then.
    (c) to help you get used to using the dipper.

(2) You should put the dipper in an upright position
the first time you start dipper training and allow the rat
to find the dipper.
    (a) true
    (b) false

(3) The light in the Skinner box should be turned on
during training.
    (a) true
    (b) false

(4) After your rat has adapted to the click of the
dipper, you may start moving it up and down more
quickly.
    (a) true
    (b) false

(5) The dipper training procedure also converts the
water into a learned reinforcer.
    (a) true
    (b) false

(6) You should click the dipper really hard. The rat
will be more likely to hear the click.
(a)   true
(b)   false

(7) How long should you raise the dipper?
(a)   1-2 seconds
(b)   3-4 seconds
(c)   5-6 seconds
(d)   until the rat stops licking it

(8) Do not raise the dipper if the rat is standing
stationary over it.
(a)   true
(b)   false

YOUR LAB REPORT

The lab reports consist of three parts: data, graphs, and questions. **All of these are provided in this book and must be shown to your lab instructor for his or her signature after completing an experiment.**

As the experiment progresses, record your data directly on the data portion of the sheet; be sure to write legibly. You will also be presenting your data in graphic form. For each minute, plot the number of responses emitted during that minute. It is a good idea to use tally marks when recording data for each minute. That way there is no guessing for that minute in case you lose count.

Note the space at the top of your lab report for your instructor's signature. Always be sure that your lab instructor observes the completion of each experimental phase and signs your sheet when your rat is performing the final response criterion. (Raise your hand to get your instructor's attention right near the end.) This is very important! Also make sure the TA signs your completed report, or you will lose points.

After you have completed all of the experimental phases, it's time to graph your data. Plot the number of responses that occur in each given minute. For example, the following hypothetical data would look like this:

Minute	No. of Responses
1	3
2	5
3	7
4	10

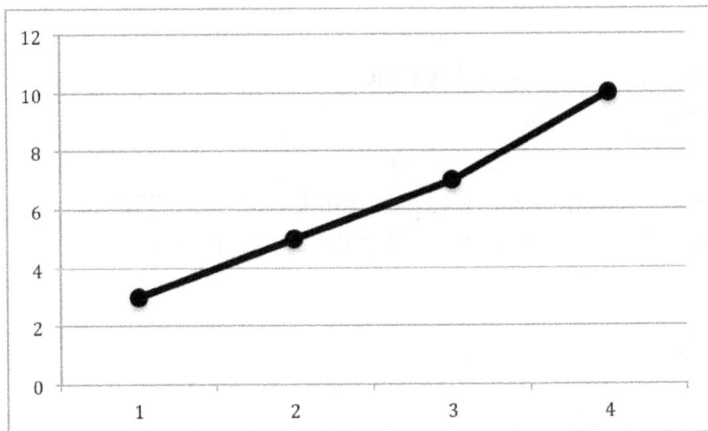

After you've plotted the points for your data, connect them with straight lines. Do not attempt to draw them free hand; use a straight edge.

Next, answer the questions for the appropriate experiment based on your data.

NOTE: Questions such as, "What minute or period had the highest response rate?" sometimes require more than one answer. For example, if the highest rate you recorded was 30 responses per minute, but this occurred three times, during the 1st, 8th, and 10th minutes, then you would write 1, 8, 10 in the appropriate blank:

Sample question: During conditioning, the highest number of responses was recorded for the   _1st, 8th, and 10th_   minute(s), the rate was   _30_   responses per minute).

So here's a general sequence you'll follow through each experiment over the semester:

THE 10 STEPS TO EXPERIMENT COMPLETION

1. Read the experiment ahead of where you are (so you should read experiment two before you finish experiment one)
2. Answer the questions at the end of each experiment's procedures.
3. Ask your TA to initial the questions so you can begin the experiment.
4. Train and take data. Ask your TA questions freely by raising your hand—they're happy to give advice, and they probably will anyway to help you along.
5. When you're near or at the end (your rat is performing at the criteria the experiment specifies), raise your hand to have your TA observe.
6. Your TA signs the top of the report when he or she sees you're cool with moving on to the next experiment.
7. Continue to the next experiment if the lab is not over.
8. Complete the graph and report outside of the lab (it's not considered participation in lab).
9. In the next lab session, show your TA your report and get his or her signature.
10. That's it! Back to #1.

EXPERIMENT I
MOVEMENT AND DIPPER CLICK TRAINING

Experimenter _____ **Date** _____ **Lab time** _____

Lab Instructor's Signature (when they observe the rat's final performance)_____

Lab Instructor's Signature (when the report is complete for the next lab)_____

Minute	No. of Responses
1	
2	
3	
4	
5	
6	
7	
8	
9	
10	

(1) How many laboratory sessions were required to complete the experiment? _____ sessions.

(2) During training, the highest number of responses was recorded for the _____ minute(s); the rate was _____ responses per minute.

(3) During training, the lowest number of responses per minute was recorded for the _____ minutes(s); the rate was _____ responses per minute.

(4) A total of _____ responses were recorded during the ten minutes of dipper click training. This was an average of _____ response per minute. (Note: be specific -- use decimals in your calculations.)

✓ **Don't forget to graph your results before the next lab!**

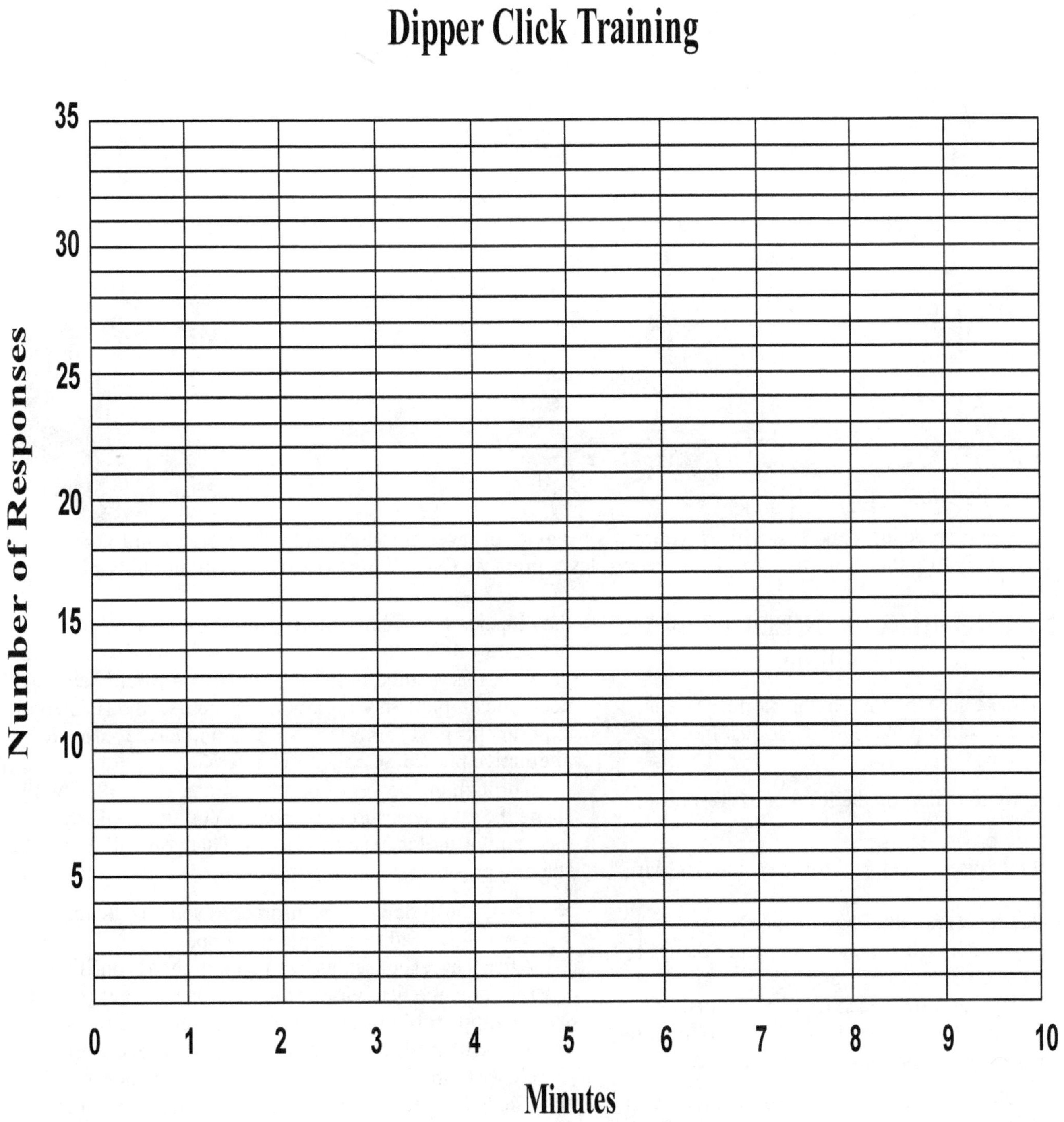
Dipper Click Training
Number of Responses
35
30
25
20
15
10
5
0
1
2
3
4
5
6
7
8
9
10
Minutes

EXPERIMENT II

Conditioning and Extinction

Follow the conditioning procedures to see your rat progress from moving his head towards the lever, to raising his head and paws above the lever, finishing with a complete lever press, shown above. (NOTE: there are many steps in between)

Follow the extinction procedures and you will see your rat's behavior of lever pressing gradually decrease in frequency, shown above. Simply turn the light off, sit back and record lever presses. (NOTE: this is after the conditioning phase)

THE CONDITIONING PROCEDURE or He was deprived, ready and willing.

Remember, to make sure the light in the Skinner box is **ON**, while you are training the lever press, just like in dipper click training.

Now you're ready to condition the lever-press response.

The conditioning procedure is:
(a) a series of exercises developed for the Canadian Air Force.
(b) best avoided by effete, intellectual snobs
(c) used by Boy Scouts when they have impure thoughts. (They sit in a tub of ice-cold water.)
(d) giving a reinforcer (click) each time the desired response occurs. Once the response occurs frequently, we say the response has been *conditioned*.

If you're having trouble answering this one, let me give you a hint: this laboratory experiment will consist of conditioning the rat's lever press.

Now, the simplest way to condition the lever-press response would be to place the rat in the Skinner box and wait until he presses the lever and then deliver the reinforcer (the click, followed by a little drop of water in

the dipper). If you have all day this procedure might work; but since you'll only be working a few hours a week with your rat, this method might mean that you would have to get an incomplete in your laboratory course and finish the experiments next semester. So, you will have to use an alternative procedure called **SHAPING**.

Through the power of reinforcement, you will actually shape the rat's lever press response. You will mold it, as the sculptor molds his clay. This is what you will do for the shaping procedure:

❖ First, you'll deliver a reinforcer as soon as the rat moves his head away from the dipper.
❖ After that response has been conditioned, you'll reinforce the behavior of moving his head away from the dipper only when it is in the direction of the lever.
❖ You will gradually require that the rat move closer and closer to the lever until he eventually touches it.
❖ Finally, you'll only reinforce the response of pressing the lever, and gradually insist that the lever be depressed all the way down (so it can't move any farther down) before reinforcing the response.

So, the shaping procedure consists of:
Differentially reinforcing successive approximations to the terminal behavior.

- Almost continually deliver drops of water contingent on approximations--at least 5 per minute or so.
- If you find that you are not delivering at least 5 drops per minute, contingent on some approximation to a lever press, then you might have set the criterion/standard for water delivery too high.
- In other words, keep moving up the sequence of approximations, requiring more and more of a lever-press like behavior, but don't look for too much too soon.

What do all of them five-dollar words mean you ask? Well, *differential reinforcement* is reinforcing one set of responses while extinguishing another. For example, you may differentially reinforce the rat's response of moving his head one-quarter inch toward the response lever. But, if he only moves it one-sixteenth of an inch toward the response lever, you may withhold the reinforcer.

What about *successive approximations*? After you have conditioned the rat's response of moving his head one-quarter of an inch toward the dipper, you raise your criterion for reinforcement. You now reinforce only the response of moving his head at least a half-inch toward the lever. This new response more closely resembles the TERMINAL BEHAVIOR of a complete lever press. As you gradually raise your criterion for reinforcement, the behavior that you reinforce more and more closely approximates the terminal behavior. It successively approximates the terminal behavior, lever-pressing. So, you see, you are differentially reinforcing successive approximations to lever pressing; you are shaping the terminal behavior.

Fortunately, your rat is not perfect. If he were, you'd never be able to shape a lever-press response. Suppose the first form of the response that you reinforce is movement of the rat's head one-quarter of an inch from the water dipper, and in the direction of the response lever. Now further suppose that your rat is perfect. He moves his head one quarter of an inch and you reinforce the response. He does it again, and you do it again, etc. He always moves his head exactly one-quarter of an inch, whether your reinforce it or not. You break out in tears because you have been sitting there for hours waiting for the rat to move his head one half an inch from the dipper, and he never does it. Your solicitous lab supervisor runs over to console you, but how can one really console a student who has failed to shape the lever-press response?

Fortunately, this is only a hypothetical example. It will never occur, because your rat isn't perfect. Sometimes, instead of moving his head one-quarter of an inch from the dipper, he'll overshoot; he'll move his head half an inch. And what do you do? You reinforce that response. Now, since the half-inch response has just been reinforced, it will be a little more likely to occur, perhaps not the next time, but before too long. When it does occur, you reinforce it

again. Gradually, it'll be occurring more and more frequently; but now a strange thing happens. Once in a while, the rat overshoots; he moves his head a half-inch from the dipper but keeps right on going, and Wowie! It's three-quarters of an inch from the dipper before he stops. And you're right there with another reinforcer. History repeats itself, the three-quarter inch response starts occurring more and more frequently and you reinforce it every time it does. Because there's variability in the rat's behavior, he does not always respond exactly the same way every time, and you are able to differentially reinforce successive approximations to the terminal behavior -- you are able to actually shape the terminal response.

***Shaping:** while we are calling the procedure you are using "shaping" you are really shaping up a few behaviors that are part of a serious of a behavior, a behavioral chain. More specifically you will be shaping up two different behaviors. The first behavior you are shaping is the rat facing the lever (the terminal behavior in the part of the chain). You will shape this behavior by reinforcing the rat's behavior of moving his head one-quarter inch toward the response lever and gradually requiring him to face the lever. The next behavior you will be shaping up is having two paws on the lever (the terminal behavior in this part of the chain). The rat will start with touching the lever with one paw and you will shape up his behavior until he consistently puts two paws on the lever. The last behavior in the behavioral chain is a two pawed lever press that presses the lever all of the way down (the terminal behavior). You will shape this behavior by first requiring the rat to push the lever a quarter of the way down until you shape his behavior up to pushing the lever all the way down with two paws. You will learn more about shaping in Chapter 8 and more about chaining in Chapter 20.

Ok, so now let's get a little more complicated.

IMPORTANT!
Many students screw up in rat lab because they fail to heed the following advice: Once your lever press is established, **ONLY REINFORCE LEVER PRESSES THAT GO ALL THE WAY DOWN AND ARE TWO-PAWED PRESSES!** We can't stress how important this is. If you fail to do this, you will have serious troubles with the experiments you will do. Your lab instructor will make you go back and recondition the lever press, and that will take a long time. Plus, everybody will think you are a shmuck.

It's very logical. If the rat can get the water by pressing the lever just a little, that's all he will do. You have to make him press that thing all the way down! How can you tell when your rat is pressing the lever all the way down? You should check out the lever before putting the rat in the cage to see how far it goes. (Stick your hand in the box and push down the lever as far as it will go.) In general, when the lever goes all the way down, the long screw on the other end of the Skinner box (outside of the box) will hit the back of the light.

Questions

(1) What procedure will you be using to get the rat to press the lever?

(a)   shaping

(b)   bribing

(c)   modeling

(d)   begging

(2) To shape a lever press, you simply wait as long as you have to until the rat presses the lever. Then you deliver the water.

   (a) true

   (b) false

(3) How far down does the lever have to go?

   (a) about halfway

   (b) just enough that you can see it's going down

   (c) about three-quarters of the way

   (d) all the way down baby!

(4) What will happen if you do not condition a lever press that is all the way down?

   (a) you will have trouble with all the experiments.

   (b) your lab instructor will make you condition the lever press again.

   (c) your peers will all laugh at you.

   (d) all of the above.

EXTINCTION

(5) Here's another fun technical term. To extinguish a response you:

   (a) spray it with CO2

   (b) stop presenting the reinforcer when the organism makes the response.

To extinguish your rat's lever pressing, you stop presenting the click when he presses the lever. Extinguishing a response will cause it to decrease in frequency.

When you differentially reinforce one form of a response, you are differentially extinguishing all other forms of the response. By withholding reinforcement for these other forms of the response, they will occur less and less frequently. So, shaping consists of differentially reinforcing successive approximations to the terminal response, and differentially extinguishing forms of the response that are not close enough to the terminal response.

Shaping the behavior of your own living organism is a subtle and delicate art. A common error is to raise your criterion for successive approximations in steps that are too large. Remember our motto, "think small." The skilled shaper gradually raises his or her criterion a silly millimeter at a time. Reverting to a more familiar system, suppose the rat is moving his head one-quarter of an inch toward the lever three or four times a minute and you reinforce that response whenever it occurs. In the meantime, you are waiting until he moves it one-half an inch toward the lever but he never moves it that far. Is the rat wrong? No, THE RAT IS NEVER WRONG. You're thinking too big; remember our motto. Now you need to start using your calibrated eyeball.

Surely, once in a while his quarter-inch response is a little longer than at other times. In other words, occasionally he moves his head 9/32 of an inch, 1/32 of an inch longer than your criterion. Then let that be your new criterion; require 9/32 of an inch response rather than a half-inch response. Impossible, you say. You feel you can't make that sharp discrimination. Well, perhaps, but if you get your old eyeball down there where the action is and stay in a fairly constant position, picking out some reference spot on the Skinner box wall that you can measure the behavior against, you'll be in pretty good shape. Shape the response 1/32 of an inch at a time, so the animal just flows over to that lever.

The shaping procedure is actually an even more gradual process than we've implied. Suppose you've been reinforcing the 9/32 of an inch response and now he's occasionally making 10/32 of an inch response as well, moving 10/32 of an inch from the lever so of course you begin reinforcing that response. If the 10/32 of an inch response is only occurring once every three minutes, you may wish to continue occasionally reinforcing the 9/32 of an inch response. If you don't reinforce an average rate of at least once or twice a minute, the behavior may extinguish altogether and the animal will go to sleep in the corner. So gradually shift from one criterion to the next slightly higher one, being careful not to do it too abruptly.

Here's another problem: don't condition the 10/32 of an inch response so firmly that the rat practically never makes the 11/32 of an inch response. If one form of the behavior becomes conditioned too strongly (because it has received too many reinforcers) the behavior may never advance to a slightly improved variation which you can differentially reinforce. If you have over-conditioned one stage of the response, then you might move from continuous reinforcement to intermittent (occasional) reinforcement. This may increase the variability in the animal's behavior so occasionally he'll emit the 11/32 of an inch response. At the same time, you may need to get your eyeball calibrated more finely so you can distinguish between a 20/64-inch response and a 21/64-inch response. More realistically, you might look at the rat's overall behavior and notice that some 10/32 of an inch responses involve just the extension of his head, whereas other times he seems to shift his entire body closer to the lever. You then reinforce only those responses involving the shifting of the whole body. If you increase the likelihood of "whole body" 10/32-inch responses, it may be that he'll occasionally stumble a little farther and give you the desired 11/32 of an inch response. Always look closely at the behavior so you can anticipate the variation of the response to differentially reinforce next.

It may take several days to shape the lever-press response. If so, you might find at the beginning of the next day's experimental session that the animal's performance will not be quite what it was the previous day. It will then be all right to relax your response criterion a bit and start with one of your earlier successive approximations. You should have him back to the stage where you left off on the previous day within half-a-dozen reinforcers.

I remember the case of Lethargic Leonard. He waited for the rat to move one-quarter of an inch toward the lever. When that happened, Leonard put down his pencil, moved his hand over to the water dipper and delivered the reinforcer. Unfortunately, nearly one second had elapsed between the time the rat responded and the delivery of the reinforcer. That one second was sufficient time for the rat to move back toward the dipper and start scratching his side with his hind leg.

(6) Which response do you think was strengthened the most by Lethargic Leonard?
    (a) the quarter-inch move
    (b) scratching

Lethargic Leonard's rat soon became known as Raw Richard because he had developed this strange scratching response that wiped out his formerly luxurious coat. VERY IMPORTANT MORAL: Lay those reinforcers on your rat's behavior with lightening-like speed, or you'll end up reinforcing the wrong response. A good way to do this is to keep your hand on the dipper at all times – then you'll always be ready.

Remember, THERE'S NO SUCH THING AS A DUMB RAT. Most often, if the rat's lever-press response has not been conditioned within a few experimental sessions, it's because the experimenter has not been observing the behavior carefully enough, and/or has been too slow in delivering those reinforcers. But, of course, there are other ways you can screw up.

Take, for example, case no. 74 -- J3, Clem the Clod. After ten sessions of shaping, the rat finally pressed the lever. Clem found the rat's response so reinforcing that he jumped out of his seat, shouted "Hoo Ha, boy," pounded his lab partner on the back and then said "Oh yeah," and reinforced whatever the rat was doing at that time. It happened that the rat was freezing in the corner because of Clem's cloddish noises, but that rat wasn't alone. All of the other rats in the lab were cowering in the corner of their Skinner boxes. Clem had few friends. MORAL: the good experimenter is a quiet experimenter and a person with many friends.

SECRETS FOR SHAPING: or How to beat the dumb rat syndrome.

(1) Until the rat's moving his nose from the dipper, it's probably easiest to watch his nose (rather than his feet).

(2) Initially, pick a small response that the rat is currently making! This is the initial response. He may be moving his nose 1/16 of an inch, and he might only do it once every 2 minutes, but be sure to reinforce it whenever it happens.

(3) Again, leave the water dipper up for 3 to 4 seconds -- no more, no less.

(4) The biggest mistake is requiring the rat to progress too quickly.

(5) Concentrate on strengthening the initial response so it occurs frequently -- remember, this response might be only a 1/16-inch movement of the rat's nose.

(6) When the initial response is occurring once every 5 or 6 seconds, introduce a short period of extinction (don't reinforce the initial response). The rat may make the initial response a couple times, then he may move his nose farther -- reinforce when the response magnitude increases. Don't expect too much of an increase!

(7) Now reinforce any response that is closer to the lever than the initial response. The rat may move his nose 1 inch -- reinforce it, but don't expect him to do this all the time. You should condition movements approximately 1/16 or 1/8 of an inch beyond the initial behavior (so it occurs once every 5 or 6 seconds).

(8) Once this response is conditioned, introduce extinction again, then reinforce when the response magnitude increases.

(9) Continue alternating periods of reinforcement and extinction to advance the rat's behavior (in steps of 1/8 inch or so) until his nose goes by the dipper very consistently.

(10)   Now you must get the rat's nose above the lever, and his feet off the floor. When the rat's nose is by the lever, reinforce the response only if he raises his head. Wait him out -- don't give up -- he WILL raise his head -- remember, he has been deprived of water for 23 hours.

(11)   Condition head raising above the lever the same way you got him over to the lever -- alternate periods of reinforcement and extinction. Remember, this extinction period may be *very* brief -- just until the behavior advances.

(12)   Now that his head is above the lever, start watching his feet. After all, to press the lever, his feet must come off the floor. Now only reinforce responses that involve foot movement.

(13)   Advance foot movement until the rat touches the lever (by alternating conditioning and extinction periods. He may push the lever if you have already conditioned touching it. Just withhold reinforcement until he does! But, don't expect him to push it too far.

(14)   When you have a high frequency of lever pressing, start reinforcing only complete lever presses (all the way down). Once you have a high frequency of complete lever presses, **NEVER** AGAIN REINFORCE A PARTIAL LEVER PRESS! REINFORCE **ONLY** FULL LEVER AND TWO-PAWED PRESSES! Notice this isn't the first time we've mentioned this?

 To get your partner (the rat) to perform the way *you* want him to, here are some things to remember:

(1) If the rat roams around the box, WAIT! He'll return sooner or later, and you'll not have reinforced his roaming around the box by clicking the dipper until he comes over. Just WAIT!!!!! WAIT, WAIT, WAIT! (Yes, this is very important.)

(2) NEVER, NEVER, NEVER pick up the rat and set him on the lever or dipper. It will do more harm than anything else.

(3) When in doubt, wait.

(4) There is NO SUCH THING AS A DUMB RAT, just inadequate experimental techniques. If your rat is not doing what you want him to do, **it's not the rat**, it's YOU.

The following are 10 general laws of shaping found in Karen Pryor's book "Don't Shoot the Dog". She is a world-renowned dolphin trainer and if these principles work for dolphins they sure as heck better work for rats. Not all of these laws may be applicable to what you're doing right now but most of them reflect what we've just talked about in the previous paragraphs. Those that don't are still important however and you'll definitely want to refer back to them when it comes time to do your own experiment.

(1) Raise criteria in increments small enough so that the subject always has a realistic chance for reinforcement.

(2) Train one thing at a time; don't try to shape for two criteria simultaneously.

(3) Always put the current level of response onto a variable level of reinforcement before adding or raising criteria.

(4) When adding a new criterion, temporarily relax the old ones.

(5) Stay ahead of your subject: Plan your shaping program completely so that if the subject makes sudden progress, you are aware of what to reinforce next.

(6) Don't change trainers in midstream; you can have several trainers per trainee but stick to one shaper per behavior.

(7) If one shaping procedure is not eliciting progress, find another; there are as many ways to get behavior as there are trainers.

(8) Don't interrupt a training session gratuitously; that constitutes a punishment.

(9) If behavior deteriorates, "go back to kindergarten;" quickly review the whole shaping process with a series of easy reinforcements.

(10)   End each session on a high note, if possible, but in any case quit while you're ahead.

QUESTIONS

(1) The differential reinforcement procedure means that you will reinforce one response and _____ a similar, but different, response.
(a)   punish
(b)   reinforce
(c)   extinguish

(2) The extinction procedure will _____ the frequency of the response.
    (a) increase

(b) decrease

(3) If the rat is not performing as well as he should be, you probably just have a dumb rat.

(a)   true

(b)   false

(4) If the rat spends a lot of time on the side of the box opposite the lever, you should click the dipper repeatedly to get his attention.

(a)   true

(b)   false

(5) If the rat spends a lot of time on the side of the cage opposite the lever, you should wait him out (wait till he walks closer to the lever).

(a)   true

(b)   false

(6) You should raise your requirements for successive approximations in very large steps.

(a)   true

(b)   false

(7) What should you do if you have over-conditioned one stage of the response?

(a)   you should dipper train the rat again.
(b) you should reinforce *everything* except that stage of the response.
    (a) you should move from continuous reinforcement to intermittent (occasional) reinforcement.

(8) Reinforce the responses *immediately*.

(a)   true

(b)   false

There are two phases to this experiment: conditioning and extinction.

CONDITIONING (Phase I): Naked rats and humans performing unnatural acts.

After you have successfully shaped your rat's lever-press response in the Skinner box, you are ready to record data. For twenty (20) minutes, reinforce every lever-press the rat completes. Record the number of lever-pressing responses made in each minute on the supplied data sheets. Remember the light should be ON during this phase.

For example, if the rat made 6 responses in the first minute, 12 in the second, 8 in the third, and 5 in the fourth minute, your data sheet should look like this:

Minute	No. of Responses
1	6
2	12
3	8
4	5

After you have recorded data for twenty minutes, you are finished with the first phase of the experiment.

EXTINCTION (Phase II): or the rat that couldn't say no.

During the second phase, you will <u>not</u> reinforce any lever presses that the animal makes. **You should NOT begin this phase of the experiment immediately following the first phase.** Since you are using water to reinforce the animal's behavior and you have just given him water for twenty minutes, it may be the case that the water deprivation level has been greatly reduced. This means that satiation may have occurred and water would no longer be an effective reinforcer. Therefore, extinction may appear to occur more rapidly than would really be the case.

However, if you completed the first phase early in the lab period (say you've reinforced the animal's lever press for 10 minutes one day but time ran out, so you continued phase I on the following day), you can feel confident in starting the extinction phase.

If you are starting the extinction phase at the beginning of a lab period, reinforce lever pressing for four minutes before you start the extinction phase. It's important that you have some behavior to extinguish. You should record the number of responses emitted during these four minutes just as you did for phase one.

After the four minutes of reinforcement are up, do not reinforce any lever pressing. Simply turn the light **OFF** and continue to record the number of responses until the animal makes NO response for a three-minute period.

(Thus, how long this phase lasts will differ from student to student.)

One more time in case you missed it. When you start the extinction phase of this experiment, you must turn the light **OFF.**

In this phase, your primary task is to observe the decline in rate of a previously reinforced response when reinforcement is withheld. Besides the decline in rate, there is another feature of behavior, which is of importance during extinction: the resulting increase in variability of behavior. These behaviors, which replace lever pressing during extinction, may differ in form from those observed. Several investigators have reported the appearance of "aggression" during extinction. Animals sometimes bite the lever, pull on it, and so forth. Sometimes the rat may go back to doing the things he did during dipper training.

WHISTLE BLOWING: If you have not completed this lab after three lab sessions, blow the whistle! **Let your TA know so he or she can help you progress through the experiment.**

QUESTIONS FOR EXPERIMENT II

(Remember, you must answer these and have your lab instructor look at them and initial the space below BEFORE you start the experiment.)

Instructor's initials: _____

(1) How far down should you require the lever press to be?
(a)   halfway down
(b)   all the way down
(c)   one-third of the way down
(d)   just enough that the lever moves a little

(2) Before you start the extinction phase you should
(a)   reinforce the lever press for four minutes
(b)   give the rat free water access
(c)   take the rat out of the box for a few minutes

(3) In the extinction phase the light in the Skinner box should be turned
      (a) On
      (b) Off

(4) You do not have to have your lab instructor review your lab reports before going on to another experiment.
(a)   true
(b)   false

EXPERIMENT II:

CONDITIONING AND EXTINCTION

Experimenter _____ **Date** _____ **Lab time** _____

Lab Instructor's Signature (when they observe the rat's final performance)_____

Lab Instructor's Signature (when the report is complete for the next lab)_____

Conditioning

Minute	No. of Responses	Minute	No. of Responses
1		11	
2		12	
3		13	
4		14	
5		15	
6		16	
7		17	
8		18	
9		19	
10		20	

(5) How many laboratory sessions were required to complete the experiment? _____ sessions

(6) During conditioning, the highest number of responses was recorded for the _____ minute(s); the rate was _____ responses per minute.

(7) During conditioning, the lowest number of responses per minute was recorded for the _____ minutes(s); the rate was _____ responses per minute.

(8) A total of _____ responses were recorded during the twenty minutes of lever pressing. This was an average of _____ responses per minute. (Note: be specific -- use decimals in your calculations.)

Extinction (begin at a time when satiation shouldn't be a confounding variable)

Minute	No. of Responses	Minute	No. of Responses	Minute	No. of Responses
1		14		27	
2		15		28	
3		16		29	
4		17		30	
5		18		31	
6		19		32	
7		20		33	
8		21		34	
9		22		35	
10		23		36	
11		24		37	
12		25		38	
13		26		39	

(1) It took a total of _____ minutes to reach the criterion for extinction, during which _____ (how many) responses were emitted. This was an average of _____ responses per minute.

Be sure to complete the graphs on the next two pages!!!

Conditioning

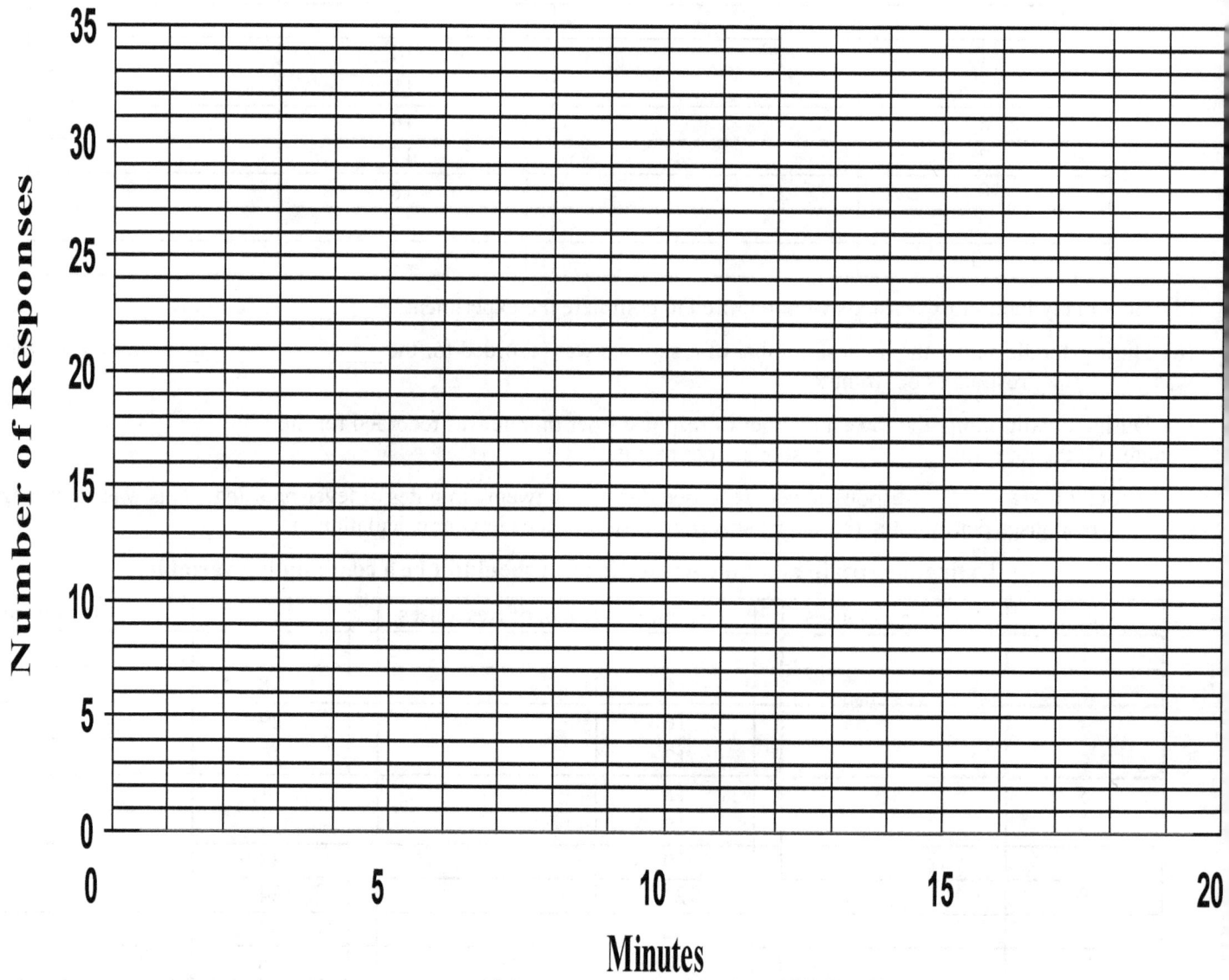

Y-axis: Number of Responses (0 to 35)
X-axis: Minutes (0 to 20)

Extinction

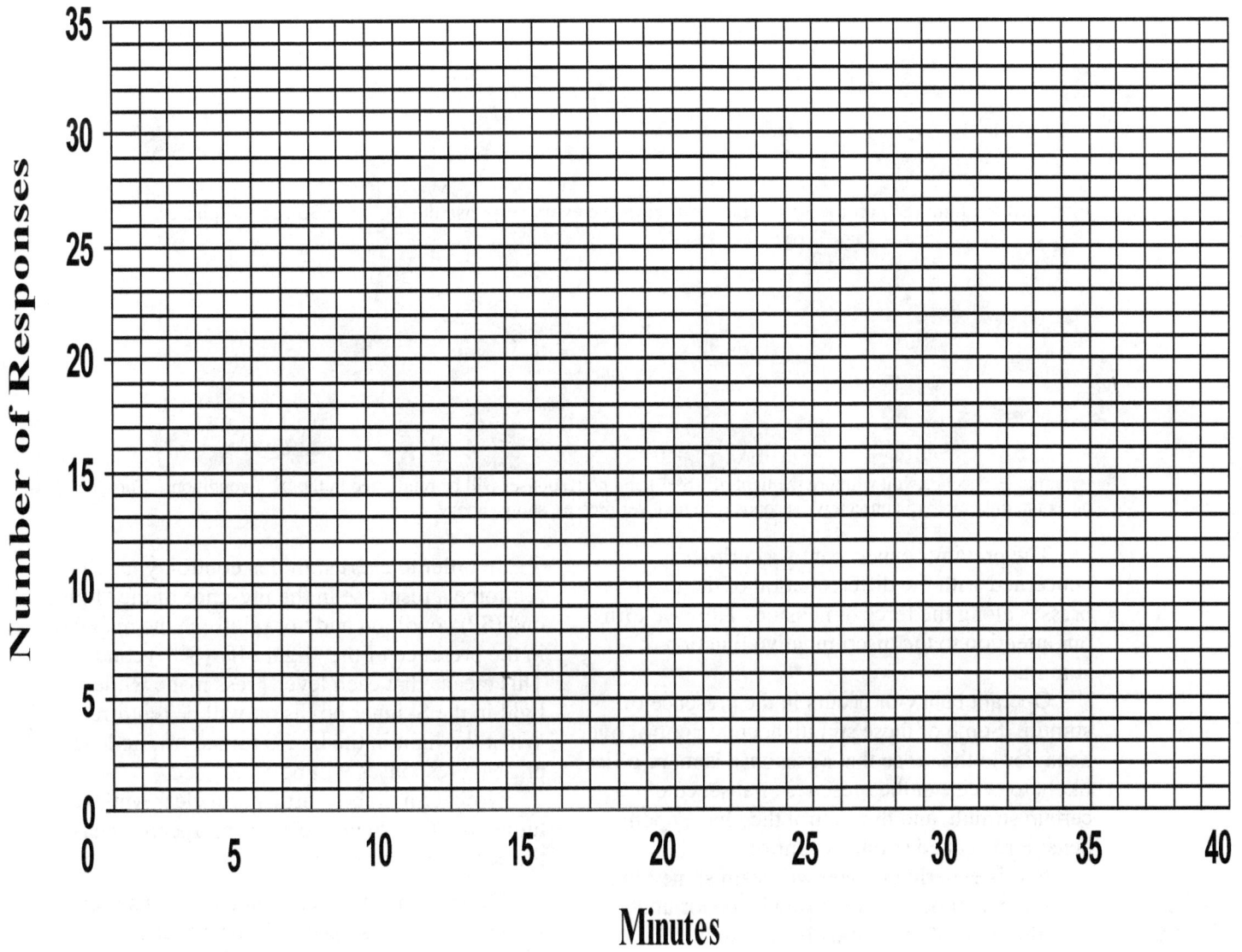

Number of Responses (y-axis, 0 to 35)

Minutes (x-axis, 0 to 40)

EXPERIMENT III

Discrimination Training: What really Goes on When the Lights Go Out

First turn the light (S^D) on for 30 seconds and reinforce every response during these 30-second phases.

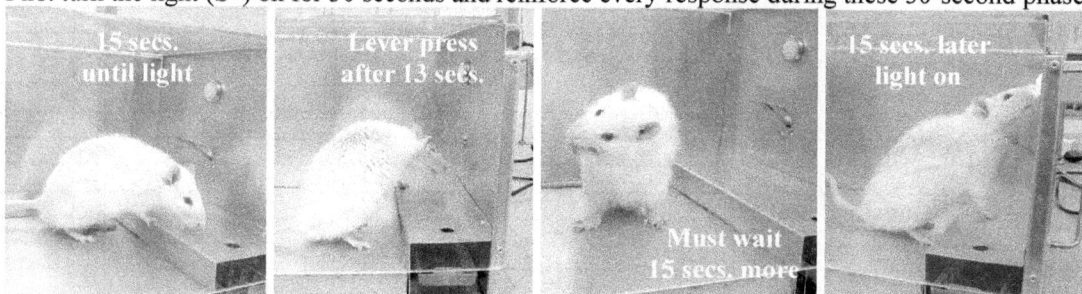

After the 30 seconds, you will turn the light off (S^{delta}) and no responses will be reinforced in the S^{delta} conditions. Turn the light on again, **only** after 15 consecutive seconds with no lever presses, shown above.

The previous experiment was primarily concerned with the differentiation of the lever presses along the force dimension. Now we turn our attention to the discriminative function of stimuli.

Operant behavior occurs in the presence of stimuli. Some of these stimuli acquire control over behavior in the sense that a response is more or less likely to occur, in the presence or absence of certain stimuli, and the control they have upon behavior is called stimulus control.

In this experiment, you will learn something about the methods used to bring behavior under stimulus control. Specifically, you will

(a) acquire control over lever pressing with a stimulus which, in the beginning, exerts very little control and

(b) observe the systematic growth of this control.

In order to exert stimulus control, you will reinforce a response in the presence of the "light on" (S^D) condition and not reinforce the response in the presence of the "light off" (S^{delta}) condition. This means that each lever-press made while the light in the Skinner box is on will be reinforced. When the light in the box is turned off, no lever-pressing responses will be reinforced.

Now it might be a little more clear why we turned the light on and off during specific phases in the earlier experiments.

You will need either a watch with a sweeping second hand or a stop-watch for this experiment. The duration of the "light on" (S^D) periods will be 30 seconds and the duration of the "light off" (S^{delta}) periods will vary depending on when the responses occur.

This is what you should do:

Turn the light (S^D) on in the Skinner box, and keep it on for 30 seconds. Reinforce every complete and two-pawed lever press the rat makes during these 30 seconds and record how many responses he makes on the data sheet provided. After 30 seconds has elapsed, turn the light off.

When the light is off, do not reinforce any lever presses. **Once the light is off (S^{delta}), 15 consecutive seconds of no responding must elapse before the light is turned on again.** For example, you might start an S^{delta} period and observe that no responses occur during the first eight (8) seconds. Then a response occurs in the ninth second. Therefore, you must wait 15 more seconds for a total of 24 (9 + 15) seconds. If no responses occur during the remaining 15 seconds, then you can turn the light back on again and start the next S^D period.

On the other hand, it might go like this: you start the S^{delta} period and the rat responds in the ninth second, so you start timing a new 15-second period. If all goes well, you'll be able to turn the light on after a total of 24 seconds. However, all does not go well; during the 23rd second, the rat responds again so you and the rat have at least 15 more seconds to spend in the dark. Now if your rat is cool, he'll have a total of 38 (23 + 15) seconds in the dark. And if he's not cool, you'll have to add another 15-second period after the next appropriate response.

Suppose the rat was really into responding and always responded at least once every 14 seconds. How long would you stay in S^{-delta} before you turned on the ($S^{D)}$) light? Right you are. You'd be well into the Dark Ages; that light would never come back on. The light stays off *until 15 consecutive seconds pass without a single lever-press response.*

Suppose that after 30 seconds of S^D pass with the light on, regardless of whether or not the rat responds, you then shut the light off and go into S^{delta}. After 13 seconds in S^{delta}, the rat presses the lever once and then responds no more. How long will the total S^{-delta} period be?
(a) 15 seconds
(b) 30 seconds (two 15 second S^{delta} periods)
(c) 28 seconds (13 + 15)
(d) none of the above

As you no doubt realize, the answer is "c", 28 seconds. If you put "a" (15 seconds), you were just kidding us, weren't you? If you put "b" (30 seconds), you should realize that the second 15-second period starts as soon as the response occurs; you don't wait until the first 15-second period has elapsed.

Continue alternating these stimulus conditions *until three consecutive S^{delta} periods have occurred in which no responses were emitted AND at least one response occurred in each of the intervening S^D periods.* The last entries on your data sheet should look like this:

Period	No. of responses in S^D	No. of Responses in S^{-delta}
95	1 (or more)	0
96	1 (or more)	0
97	1 (or more)	0

When you graph the S^{delta} periods, remember that the length of the periods will vary. They should be 15 seconds only after the discrimination has been established. When you start with the first S^{-delta} period, probably a response will occur within the first 15 seconds.

The S^D periods will always be 30 seconds in length, while the length of the S^{-delta} periods will vary.

When recording your data, you should record the S^D and S^{delta} periods separately. In each case, record the number of responses emitted in the S^D and S^{delta} periods. (see the table below.)

Period	No. of responses in S^D	No. of responses in S^{-delta}
1	18	5
2	20	3
3	5	6
4	17	2
5	13	1
6	17	0

The data for both the S^D and S^{-delta} periods should be plotted on the same graph. The S^D points are designated by circles (i.e. O--O--O), and the S^{-delta} points are designated by crosses (i.e., X--X--X). For example, the above data would be graphed as follows:

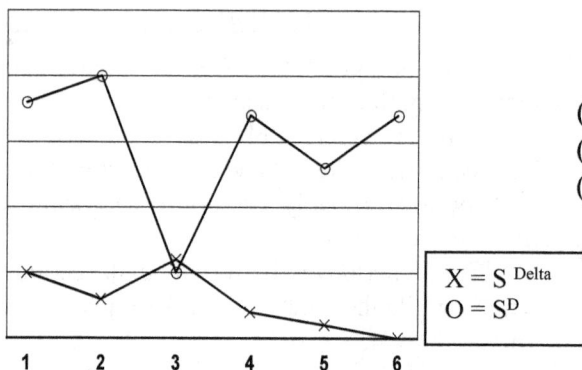

X = S ^{Delta}
O = S^D

1   2   3   4   5   6

IMPORTANT: On your graph, plot only the FIRST and the LAST twenty-five (25) periods!

WHISTLE BLOWING: If you have not completed this lab after three lab sessions, blow the whistle! Let your TA know so he or she can help you progress through the experiment.

QUESTIONS FOR EXPERIMENT III

(Remember, you must answer these and have your lab instructor look at them and initial the space below BEFORE you start the experiment.) Instructor's initials: _____

(1) What condition is "light on"?
(a) S^D
(b) S^{delta}

(2) In the presence of the light, the response will be _____.
(a)   reinforced
(b)   extinguished

(3) In the absence of the light, the response will be _____.
(a)   reinforced
(b)   extinguished

(4) During the S^{-delta} condition, your rat presses the lever after 9 seconds. How many *more* seconds will the condition last if the rat does NOT press the lever again?
(a)   6 more seconds
(b)   15 more seconds
(c)   24 more seconds

EXPERIMENT III:

Discrimination Training

Experimenter _____ **Date** _____ **Lab time** _____

Lab Instructor's Signature (when he/she has observed the rat's final performance)_____

Lab Instructor's Signature (when the report is complete for the next lab)_____

Period	No. of Responses in S^D	No. of Responses in S^{delta}	Period	No. of Responses in S^D	No. of Response in S^{delta}
1			26		
2			27		
3			28		
4			29		
5			30		
6			31		
7			32		
8			33		
9			34		
10			35		
11			36		
12			37		
13			38		
14			39		
15			40		
16			41		
17			42		
18			43		
19			44		
20			45		
21			46		
22			47		
23			48		
24			49		
25			50		

Period	No. of Responses in S^D	No. of Responses in S^{-delta}	Period	No. of Responses in S^D	No. of Response in S^{-delta}
51			76		
52			77		
53			78		
54			79		
55			80		
56			81		
57			82		
58			83		
59			84		
60			85		
61			86		
62			87		
63			88		
64			89		
65			90		
66			91		
67			92		
68			93		
69			94		
70			95		
71			96		
72			97		
73			98		
74			99		
75			100		

Period	No. of Responses in S^D	No. of Responses in S^{delta}	Period	No. of Responses in S^D	No. of Response in S^{delta}
101			126		
102			127		
103			128		
104			129		
105			130		
106			131		
107			132		
108			133		
109			134		
110			135		
111			136		
112			137		
113			138		
114			139		
115			140		
116			141		
117			142		
118			143		
119			144		
120			145		
121			146		
122			147		
123			148		
124			149		
125			150		

Discrimination Training

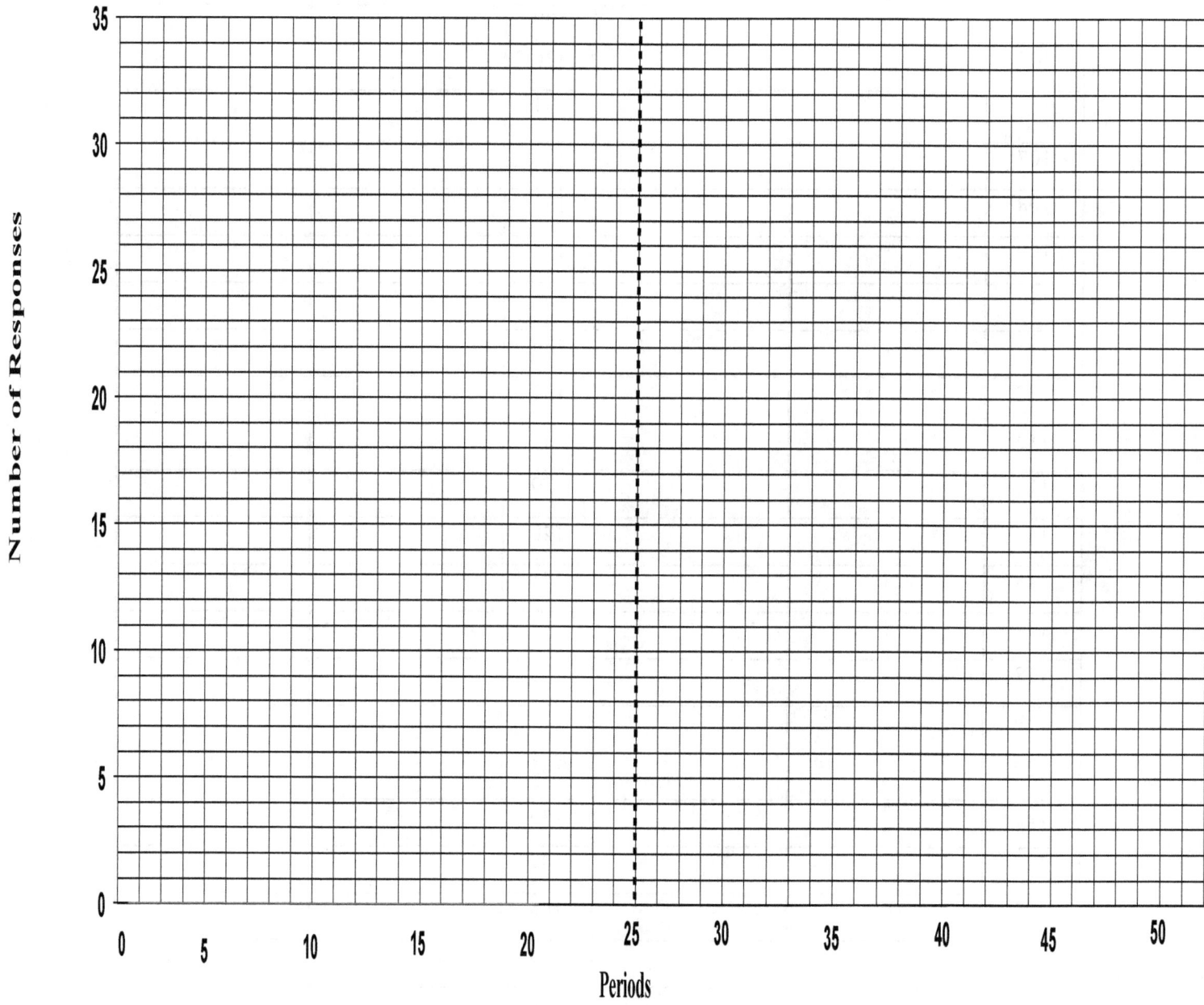

Number of Responses (y-axis, scale 0 to 35)

Periods (x-axis, scale 0 to 50)

Lab Questions

(1) How many laboratory sessions were required to complete this experiment? _____ sessions

(2) During the S^D periods, the highest number of responses occurred in the _____ S^D period(s); while the lowest number occurred in the _____ S^D period.

(3) A total of _____ responses were recorded during the S^D periods; which was an average of _____ responses per period.

(4) During the S^{-delta} periods, the highest number of responses occurred in the _____ S^{-delta} period(s); Zero responses during the S^{-delta} periods first occurred during the _____ period.

(5) A total of _____ responses were emitted during the S^{delta} periods, which was an average of _____ responses per period.

(6) How many S^{-delta} periods were required before the discrimination criterion was met? _____ S^{-delta} periods.

(7) A response is _____ in the presence of an S^D.
   (a) reinforced
   (b) extinguished

(8) A response is _____ in the presence of an S^{-delta}.
   (a)   reinforced
   (b)   extinguished

33

EXPERIMENT IV:

Stimulus-Response Chaining: *Subject in Chains*

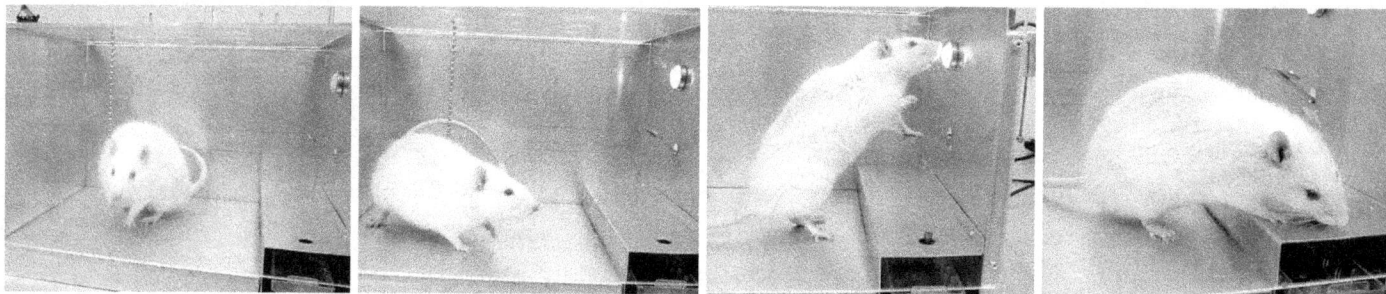

At the beginning of this experiment you will turn on the light when your rat moves toward the chain (which is an approximation response of actually pulling the chain). Once that light comes on, he'll take it from there, shown above.

Next you will turn on the light for getting closer to the chain, (standing next to it as shown above). Then, for touching the chain. Finally, you will turn the light on only when he pulls the chain, which will be followed by a lever press and water.

We have examined both the reinforcing and discriminative functions of stimuli in the previous studies. Now we are ready to observe stimulus-response chaining, in which a single stimulus serves both as a reinforcer and an S^D.

The two-component chain of behavior you will establish involves lever pressing and key-chain pulling. (When we speak of inserting the chain into the experimental chamber, the chain is a key chain and not to be confused with a chain of responses.)

This is what is going to happen: You will put a key chain about ¾ of the way through the top of the holes in the Skinner box (so most of the chain is hanging in the Skinner box). The rat will pull the chain down through the holes until it stops at the clip. When the rat pulls the chain, you will turn on the light. In the presence of the light, the rat will press the lever and you will deliver the water. Then the chain of responses will start over again.

At the beginning of the first session, stabilize the lever-press behavior in the presence of the light for three minutes on a continuous reinforcement schedule (reinforce every complete lever press). Then turn the light off. NOTE: The three minutes of continuous reinforcement is a one-shot deal. ONLY do it during the

very first session. Jump right into stimulus response training for all of the following sessions

You will then want to shape the key-chain pulling response (with the light off). You will do this in the same way that you shaped lever pressing.
First, reinforce:
(a) turning toward the key chain
(b) approaching the key chain
(c) touching the key chain (knocking it, playing with it, etc. even if it is "by accident"),
(d) grasping or holding the chain, and finally
(e) pulling the chain down.

You will reinforce the key-chain pulling response by TURNING THE LIGHT ON. Why does the light function as a reinforcer? Simple. The light functions as a reinforcer because in experiment IV you reinforced lever presses in the presence of the light, but not in the absence.

So, when your rat touches (or approximates touching) the key chain, you immediately turn on the light. Then wait for the rat to press the lever. Reinforce the lever-press with the dipper click and turn off the light. Repeat this procedure until the entire stimulus-response chain is stable (four to six responses per minute), and then record responses for ten minutes. One complete chain of responses (chain-pull followed by lever-press) will be recorded as one response. After you record your responses (be sure to have your lab instructor observe the chain-pull and sign your sheet), make the graph and answer the lab questions.

In this experiment, the light serves two functions:
1) as a reinforcer for the chain pull
2) as an S^D for the lever press (lever presses are reinforced in its presence and not in its absence)

In case you're not progressing in the shaping process: You may be turning the light on perfectly, after successive approximations to the chain pull, but if your rat does not quickly press the lever after you turn on the light, you need to return to the first step of training. You test the latency of the response by turning on the light and then observing how long it takes for him to press the lever. Once the light comes on, the rat must press the lever within 3 seconds. If he fails to do this, you must train the *light on* S^D more—by going back to Experiment IV's discrimination training. If it presses the lever within 3 seconds of the light coming on, 10 times in a row, then he is ready for this experiment.

WHISTLE BLOWING: If you have not completed this lab after two lab sessions, blow the whistle! Let your TA know so he or she can help you progress through the experiment.

QUESTIONS FOR EXPERIMENT IV

(Remember, you must answer these and have your lab instructor look at them and initial the space below BEFORE you start the experiment.)

Instructor's initials: _____

(1) What is the very first response that you will reinforce by turning the light on, at the beginning of this experiment?
(a) the lever press
(b) turning towards the chain
(c) touching the chain
(d) pulling the chain

(2) What will you use to reinforce the chain pull?
(a)   turning on the light
(b)   the click
(c)   the water
(d)   taking the chain out of the chamber

(3) What will you use to reinforce the lever press?
(a)   turning on the light
(b)   putting the chain in the chamber
(c)   the click
(d)   taking the chain out of the chamber

(4) In this experiment, the light functions as a/an _____ for the chain pull and a/an _____ for the lever press.
(a)   reinforcer
(b)   stimulus-response chain
(c) S^D

(5) During the entire chaining experiment, how many times will you have given three minutes of continuous reinforcement?
(a) one
(b) two
(c) four
(d) depends on the rat

EXPERIMENT IV:
STIMULUS-RESPONSE CHAINING: Subject in Chains

Experimenter _____ **Date** _____ **Lab time** _____

Lab Instructor's Signature (when he/she has observed the rat's final performance)_____

Lab Instructor's Signature (when the report is complete for the next lab)_____

Minute	No. of Responses
1	
2	
3	
4	
5	
6	
7	
8	
9	
10	

(1) How many laboratory sessions (periods) were required to complete the experiment? _____ sessions

(2) The highest number of responses occurred in the _____ minute(s). The lowest number of responses occurred in the _____ minute(s).

(3) A total of _____ responses were recorded during the ten minutes. This was an average of _____ responses per minute.

(4) In a stimulus-response chain, each stimulus of the chain may serve as both a/n _____ for the response that produced it, and a _____ for the following response.

(5) In this experiment, the light functions as a _____ for chain pulling and a _____ for lever pressing.

DON'T FORGET TO PLOT YOUR DATA ON THE GRAPH! ☺

Stimulus-Response Chaining

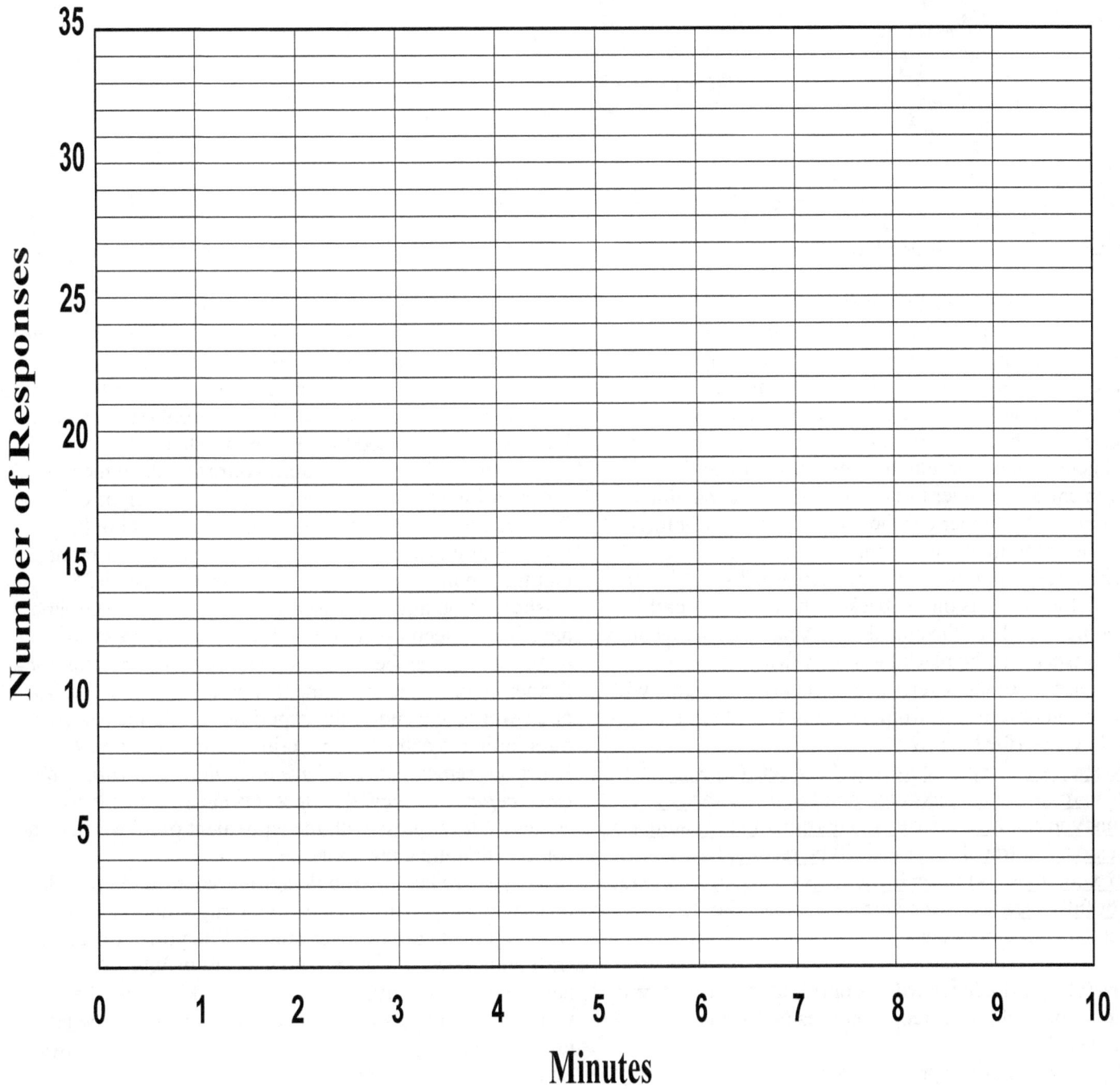

Number of Responses (y-axis: 0, 5, 10, 15, 20, 25, 30, 35)

Minutes (x-axis: 0, 1, 2, 3, 4, 5, 6, 7, 8, 9, 10)

EXPERIMENT V:
Weight Training

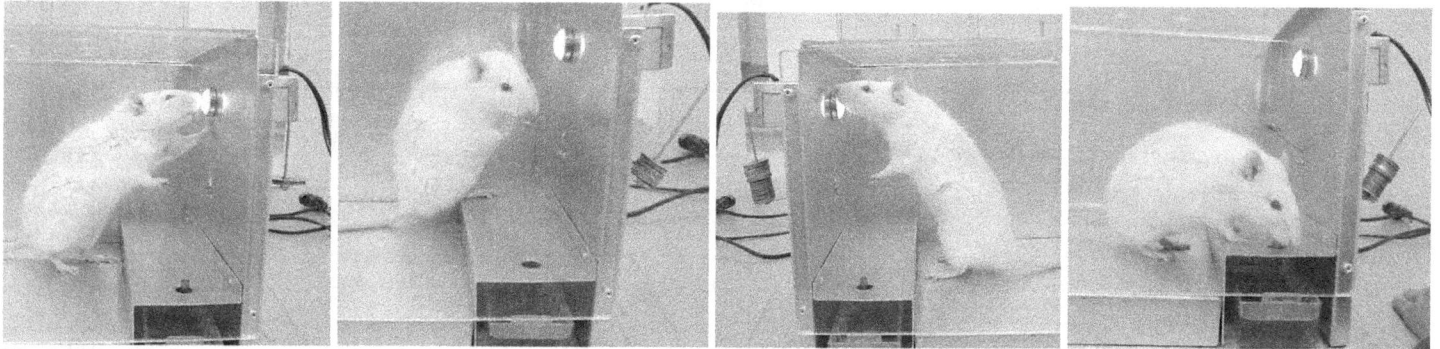

Your rat will show the strength of 10 mice as you increase the weight, one washer at a time, reinforcing each lever press.

At this point in the laboratory activities, your rat should have a well developed lever-press under the control of water deprivation (the relevant establishing operation). If the force required to press the lever suddenly became much greater, the rat would probably not learn to press with the needed force, and lever pressing would extinguish. In this experiment, you will gradually increase the force requirement on the lever to the point where the rat either cannot press any harder or where the amount of reinforcer is insufficient to maintain the more effortful response.

The force requirement will be increased by manipulating weights on the back of the lever (the part protruding outside of the chamber). A machine screw onto which washers can be placed has been fastened to the outside portion of the lever. The washers and weights will be placed on or removed from the lever according to the procedure described below.

Responses will be recorded as "correct" (+) when the lever is pressed to its maximum point of downward travel - - so that you can hear it strike the portion of the chamber that restricts its travel. Presses will be recorded as "approximations" (-) when the lever is contacted by the rat but not fully pressed. Each correct response should be reinforced with dipper click.

The weight on the lever will be manipulated as follows:
(1) Four consecutive correct responses (4 + signs in a row): Add 1 washer
(2) Three consecutive approximations (3 - signs in a row): Remove 1 washer
(3) Five approximations before four consecutive correct responses: Remove 1 washer.
Examples of these manipulations:

(1) ++++   Add a washer

(2) ---   Remove a washer

(3) +--+--+-   Remove a washer

As shown above, the weight will be increased only when four consecutive correct responses occur. But it will be decreased if three consecutive approximations occur, or if five approximations are made before the conditions for an increase occur. In each case, the weight on the lever is increased or decreased by only one washer.

Continue this program until your rat fails to meet the criterion for an increase at some particular weight value on five occasions, and you have achieved a weight of 25 washers or more. For example, suppose that your rat has met the requirements so that the present weight is 35 washers. You have increased it four times and each time have had to decrease it after too many approximation responses. Upon the fifth failure to meet the criterion for a further increase the experiment should be terminated. In other words, the end of the experiment is defined as your rat's failing to achieve some criterion for the fifth time. However, remember, *you have to achieve 25 washers or more before this applies*. Most rats have been able to achieve a lever press with the equivalent of 35 washers, and some with quite a few more.

If you are having trouble getting your rat to press the lever with very many washers, you might try working on shaping up a two-paw lever press. When the rat presses with both paws, it is able to press more weight. So if you have major difficulty, you can stop working with the weights and spend some time shaping up a two-paw lever-press. When your rat is reliably pressing with two paws, return to the weight training.

Before terminating the experiment, you must demonstrate your rat's performance at maximum weight. If this performance is too unstable to demonstrate, you can demonstrate the performance at the weight level achieved immediately prior to achieving the maximum level. Your instructor must sign your data sheet, giving you permission to terminate the experiment.

Record all the responses that occur during the same minute on one horizontal line and then go to the next line after 60 seconds have elapsed. When you begin a new line, record the number of washers in use at the beginning of that minute.

WHISTLE BLOWING: If you have not completed this lab after three lab sessions, blow the whistle! Let your TA know so he or she can help you progress through the experiment.

QUESTIONS FOR EXPERIMENT V

(Remember, you must answer these and have your lab instructor look at them and initial the space below BEFORE you start the experiment.)

Instructor's initials: _____

1. If the rat presses the lever all the way down 3 times then presses it ¾ of the way down on the next try, then presses it all the way down, do you
   (a) add a washer
   (b) take off a washer
   (c) do nothing

2. If the rat presses the lever two times in a row, but not quite all the way down, then he presses it four times all the way down you should
   (a) add a washer
   (b) take a washer off
   (c) do nothing

EXPERIMENT V
WEIGHT TRAINING

Experimenter _____ **Date** _____ **Lab time** _____

Lab Instructor's Signature (when he/she has observed the rat's final performance)_____

Lab Instructor's Signature (when the report is complete at the next lab)_____

Min	# of washers	Responses	Min	# of washers	Responses
1			26		
2			27		
3			28		
4			29		
5			30		
6			31		
7			32		
8			33		
9			34		
10			35		
11			36		
12			37		
13			38		
14			39		
15			40		
16			41		
17			42		
18			43		
19			44		
20			45		
21			46		
22			47		
23			48		
24			49		
25			50		

Min.	# of washers	Responses	Min.	# of washers	Responses
51			76		
52			77		
53			78		
54			79		
55			80		
56			81		
57			82		
58			83		
59			84		
60			85		
61			86		
62			87		
63			88		
64			89		
65			90		
66			91		
67			92		
68			93		
69			94		
70			95		
71			96		
72			97		
73			98		
74			99		
75			100		

1. During training, the highest number of responses was recorded for the _____ minute(s); the rate was_____ responses per minute. The experiment took _____ sessions to complete.

2. The maximum number of washers obtained during training was _____.

EXPERIMENT VI

Fixed-Ratio Schedule of Reinforcement: *How I Went All the Way* (to FR 8)!

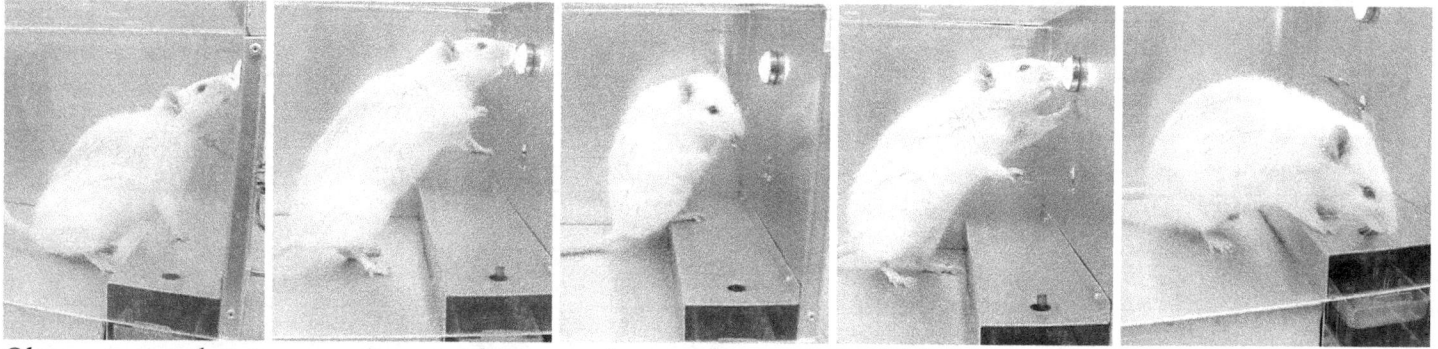

Okay rats, get those paws up, one and press and two and press now drink.

The previous experiments have demonstrated that behavior is dependent upon its environmental consequences. In the natural environment, however, the consequences of behavior are not always consistent. Behavior is not always continuously reinforced. Sometimes the reinforcement is intermittent or occasional. In the present experiment, you will use an intermittent schedule of reinforcement, one in which reinforcement only occurs after a fixed number of responses have been made. This is called a *fixed-ratio schedule of reinforcement.*

There is some terminology that you should be familiar with for this experiment. The first is *ratio run.* The ratio run begins when the rat makes his first response following reinforcement and responds until the reinforcer is delivered. For example, if you are on a fixed-ratio 4 reinforcement schedule, the ratio run is the first, second, third and fourth response, after which you will deliver the reinforcer.

The other term is *ratio strain.* Ratio strain occurs when we make the ratio too high, too quickly and the response extinguishes. In this experiment, we will say ratio strain occurs *any time there is a pause of more than 5 seconds during a ratio run.* For example, suppose your rat is on FR4. That means the rat has to press the lever four times before you deliver the reinforcer. Now, suppose your rat presses the lever 3 times. But then he scratches himself for six seconds. That is considered ratio strain. If a ratio strain occurs during this ratio run, **you would deliver the reinforcer following the response after the pause (remember this, it's a question below)**, because it is still the fourth response of that strained run. On your recording sheet circle that strained ratio run (see illustration on the next page). If five ratio strains occurred in a row, in this FR4 example, you would drop back to the FR2 level.

(1) Which of these is a ratio run?

    (a) The rat pauses more than 5 seconds between responses within a ratio run.

(b) The rat pauses more than 5 seconds between ratio runs.

(c) The time from when the rat makes the first response until the rat makes the final response and the reinforcer is delivered.

(d) Pressing the lever more than 8 times.

(2) Which of these is ratio strain?

    (a) The rat pauses more than 5 seconds between responses within a ratio run

(b) The rat pauses more than 5 seconds between ratio runs.

    (c) The time from when the rat makes the first response until the rat makes the final response and the reinforcer is delivered.

(d) Pressing the lever 8 times.

The following is a list of steps to guide you through Experiment V:

(1) Obtain a baseline of lever pressing for three minutes on a continuous reinforcement schedule. (Each response is reinforced -- it's the same as an FR1.) A "baseline" is the current rate of responding.

(2) After three minutes on FR1, increase the schedule to FR2 (every second response is reinforced). Stay on FR2 until the rat emits 25 consecutive FR2 ratio runs (Thus, a total of 50 responses, and 25 reinforcers.) without your rat pausing more than five seconds between responses in a ratio run (ratio strain). **If you do get ratio strain, you must begin counting your 25 ratio runs over again.**

(3) Once your rat has met the FR2 criterion of 25 consecutive FR2 ratio runs without ratio strain, you are ready for FR4 (the rat must press the lever four times before you deliver the reinforcer). Remain at FR4 until your friend emits at least 20 consecutive FR4 ratio runs without experiencing ratio strain. Remember, if ratio strain does occur, you must begin counting your 20 FR4 ratio runs again.

(4) Once the FR4 criterion has been met, require 15 FR6 ratio runs without ratio strain.

(5) Once this criterion has been met, require 15 consecutive FR8 ratio runs without ratio strain. After this, you will have completed Experiment VI!

Now, **if your rat shows ratio strain for FIVE ratio runs IN A ROW, you have to drop back a level**. For example, if you are on FR6 and your rat exhibits ratio strain (pausing more than 5 seconds between responses in a ratio run), you drop back to FR4 and do that level all over again!

NOTE: It's okay if a pause occurs between delivery of the reinforcer and the beginning of the next ratio run. This is called a post reinforcement pause (PRP) and is not to be confused with ratio strain, which is pausing during your ratio run.

IMPORTANT!!! There are two things to remember when doing this experiment (and any of the other experiments) in order to succeed. If you fail to do these things, *you will have an extremely difficult time with this experiment*. Do them well, and you will do well:

#1: Only count lever presses that are all the way down in your ratio run. That means if your rat presses the lever four times, but one of them was not all the way down, then you only count the rat as having pressed the lever three times.

#2: Be very quick with your reinforcer. Deliver the click *immediately* after the rat makes the final lever press in the ratio run.

You will record your data like they are shown below. Each row is a minute. Make a tally mark for each response that occurs. After CRF (3-minute baseline of continuous reinforcement), whenever you deliver a reinforcer, you record an "R" next to your tally marks. For example, if your rat is on FR4, presses the lever 4 times and you deliver the reinforcer, your data for that ratio run would look like this on your data sheet: ////R

Sometimes your ratio run will go into the next minute. That's okay, just continue making your tally marks in the next row.

Use a minus sign when you are recording to indicate between which responses ratio strain occurs. For example, if a strain occurs between the second and third response of an FR4, your data for that ratio run should look like this: //-//R

Any time your rat gets ratio strain, circle that set of tally marks. **You will have to start your required number of ratio runs at that level over again.** Notice the circled ratio strain that occurred in the eighth minute of the data below. The ratio strain occurred during the fourth FR4 attempt. Also notice, number of required FR4 ratio runs started over again at 20.

At the start of each ratio schedule (CRF, FR2, and so on) write the ratio schedule in the "ratio" column on the data sheet.

Be sure to have your lab instructor witness the experiment and sign your sheet before you stop the experiment. When this is done, answer the questions on the following page and complete the graph. In your graph, ***only plot the last 10 minutes of each ratio schedule!!!*** (make it a line graph, as with all the other experiments).

Min.	Ratio	Tally marks	Responses	Min.	Ratio	Tally marks	Responses
1	CRF	#### #### ////	14	10		//R ////R ///R ///R ///R	18
2		#### #### ///	13	11		////R ///R ///R ///R ///R	20
3		#### #### ####	15	12		///R ///R ///R ///R	16
4	FR2	//R //R //R //R //R //R //R //R	16	13		////R ///R ///R ////	16
5		//R //R //R //R //R //R //R //R	16	14	FR6	//R /////R /////R /////R /////R	26
6		//R //R //R //R //R //R //R	14	15		/////R/////R/////R/////R// ////R	30
7		//R //R	10	16		/////R/////R/////R/////R// ////R	30
8	FR4	////R ////R ////R (//-//R)	16	17	FR8	////////R////////R////////R///	27
9		////R ////R ////R //	14	18		/////R////////R////////R//////// /R…..	

WHISTLE BLOWING: If you have not completed this lab after three lab sessions, blow the whistle! Let your TA know so he or she can help you progress through the experiment.

QUESTIONS FOR EXPERIMENT VI

(Remember, you must answer these and have your lab instructor look at them and initial the space below BEFORE you start the experiment.)

Instructor's initials: _____

(1) You are on FR6 and your rat presses the lever six times and receives the reinforcer. Then your rat goes and chews on its tail for 7 seconds before starting the next ratio run. What is this an example of?
- (a) ratio strain
- (b) ratio run
- (c) post reinforcement pause
- (d) nothing

(2) You should only consider it a lever press when the lever goes ALL the way down.
- (a) true
- (b) false

(3) You are on FR4. Your rat presses the lever twice and then cleans himself for 6 seconds before pressing the lever again. What is this an example of?
- (a) ratio strain
- (b) ratio run
- (c) Post Reinforcement Pause

(4) Your rat has just completed 23 FR2 ratio runs, (only 2 more to go) and during the 24th attempt, ratio strain occurs. Do you have to go back to the beginning and complete 25 consecutive runs again?
- (a) yes
- (b) no

(5) If during an FR6 ratio run the rat pauses after the fifth response, for eight seconds, before responding again, the response that occurred after the pause would be considered.
- (a) The sixth response of that *strained* FR6 ratio run and reinforced
- (b) The first response in the next FR6 ratio run and not reinforced

(Right, the answer is *a* because we still want to reinforce the sixth response, even though the ratio run is strained and the series of runs must start again.)

(6) Your rat presses the lever all the way down 4 times and three-quarters of the way down twice. How many lever presses should you have recorded and counted?
- (a) four
- (b) six

SUMMARY OF CRITERIA FOR FIXED RATIO TRAINING:

Any time you get ratio strain, start counting your ratios over again.

If you get five ratio strains in a row, drop down to the previous level.

Ratio Level	Requirement to move onto next ratio level
CRF	Reinforce every response for three minutes
FR2	25 consecutive runs (no strain)
FR4	20 consecutive runs (no strain)
FR6	15 consecutive runs (no strain)
FR8	15 consecutive runs (no strain)

EXPERIMENT VI:

Fixed-Ratio Schedule of Reinforcement

Experimenter _____ **Date** _____ **Lab time** _____

Lab Instructor's Signature (when he/she has observed the rat's final performance)_____

Lab Instructor's Signature (when the report is complete for the next lab)_____

Min.	Ratio	Tally Marks	Responses	Min.	Ratio	Tally Marks	Responses
1				26			
2				27			
3				28			
4				29			
5				30			
6				31			
7				32			
8				33			
9				34			
10				35			
11				36			
12				37			
13				38			
14				39			
15				40			
16				41			
17				42			
18				43			
19				44			
20				45			
21				46			
22				47			
23				48			
24				49			
25				50			

Min.	Ratio	Tally Marks	Responses	Min.	Ratio	Tally Marks	Responses
51				76			
52				77			
53				78			
54				79			
55				80			
56				81			
57				82			
58				83			
59				84			
60				85			
61				86			
62				87			
63				88			
64				89			
65				90			
66				91			
67				92			
68				93			
69				94			
70				95			
71				96			
72				97			
73				98			
74				99			
75				100			

Fixed-Ratio Schedule of Reinforcement

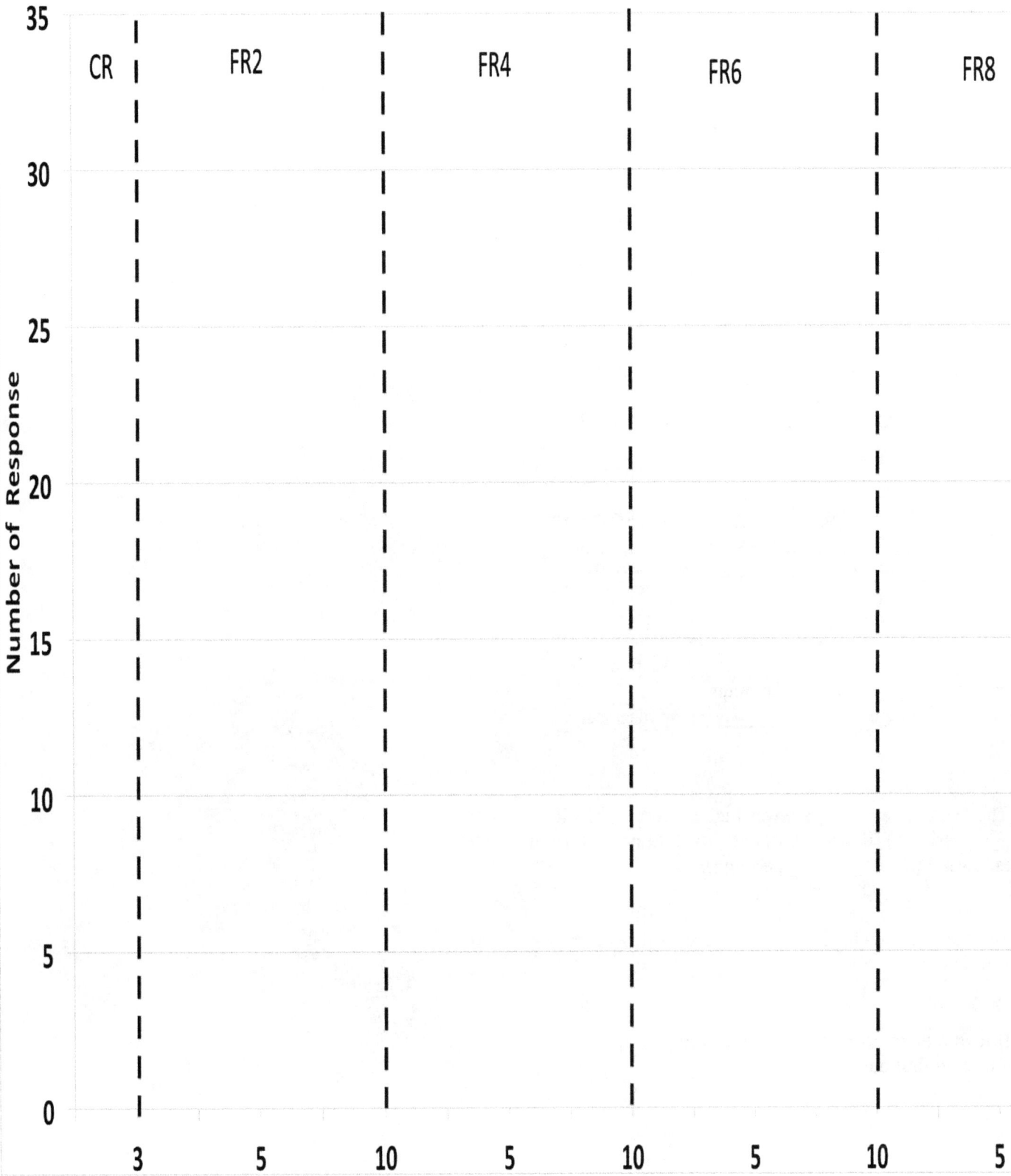

| CR | FR2 | FR4 | FR6 | FR8 |

Y-axis: Number of Response (0, 5, 10, 15, 20, 25, 30, 35)

X-axis values: 3 | 5 | 10 | 5 | 10 | 5 | 10 | 5

Questions

(1) How many laboratory sessions (periods) were required to complete the experiment?
_____ sessions

(2) During the three-minute baseline, _____ responses were recorded.
This was an average of _____ responses per minute.

(3) During FR2, a total of _____ responses were recorded for _____ minutes.
This was an average of _____ responses per minute.

(4) During FR4, a total of _____ responses were recorded for _____ minutes.
This was an average of _____ responses per minute.

(5) During FR6, a total of _____ responses were recorded for _____ minutes.
This was an average of _____ responses per minute.

(6) During FR8, a total of _____ responses were recorded for _____ minutes.
This was an average of _____ responses per minute.

(7) The average response rate was highest during the FR _____ period(s) (if it was baseline, it would be FR1), and lowest during the FR_____ period(s).

(8) Did the subject's behavior demonstrate a pause? Yes No (circle one)

(9) If he did undergo a post-reinforcement pause, at what ratio did you first observe it? _____

EXPERIMENT VII
Design Your Own Experiment

Here we are at the last experiment. Now it's time for you to show off the skills you've learned and come up with a new experiment to do.

You can use the "props" that are available at the lab or design your own. Try to come up with something that will impress the heck out of us.

IDEAS

(1) Condition this chain of responses: marble rolling (with his nose), walking through a tunnel (a piece of rolled-up cardboard or a large tube), and pressing the lever.

(2) Condition the stimulus-response chain of turning around in a clockwise direction, rolling a marble, and then turning around in a counterclockwise direction.

(3) Condition the stimulus-response chain of "chin-ups" on the wall of the Skinner box (pull himself up and then back down) and then 3 lever presses.

(4) Condition pushing a marble during "light on" and pushing a Ping-Pong ball during "light off."

(5) Condition walking to the other end of the box, walking back, and pressing the lever 2 times.

(6) Condition the rat's standing on his hind legs and turning in a clockwise circle.

(7) Condition picking up a marble and putting it on the shelf where the dipper comes out and pressing the lever.

(8) Condition jumping a small hurdle, rolling a marble and pressing the lever.

(9) Condition going through a black door during the "light on" and a white door during the "light off."

(10) Condition going down the appropriate branch of a T-maze. A T-maze consists of a straight runway, one end of which is another runway, perpendicular to the first. The rat begins at the base of the T, goes down the runway toward the top of the T, then turns one way or the other. If he turns the correct direction, the response is reinforced. (Use a large, moveable black square to indicate the appropriate branch.)

(11) Marco Tomasi, a Fall 2001 Super A student in P360, trained his rat to do an elaborate chain of behaviors. First the rat touched a marble, which would turn on a red lamp, then the rat spun a wheel, which would provide the opportunity to pull the chain, then the rat would pull the chain, which would turn on the light, and finally the rat would press the lever and drink a drop of water. Marco began the procedure with Exp. 7 and finished after completing Exp. 9. And he earned OAPs for his work developing the whole chain for his experiment 8 and 9.

(12) Ashley Fromson attempted to train his rat to sit on a chair; the only time a 360 student has tried to shape such a novel topography for a rat. His own experiment is included on the next couple pages, and he writes in his write-up how an original experiment can evolve and adapt.

Be sure to tell your lab instructor what you are going to do before starting this experiment so you don't head in a wrong direction.

WHAT NEEDS TO BE IN YOUR ROUGH DRAFT?
- **5 questions specifically relating to your experiment**
- **MUST BE TYPED**
- **NOT MORE THAN 2 PAGES**
- **MUST BE IN SAME FORMAT AS THE PREVIOUS 6 EXPERIMENTS**
- **MUST BE WRITTEN SO THAT SOMEONE ELSE COULD DO IT**
- **DO NOT USE THE WORD "I**

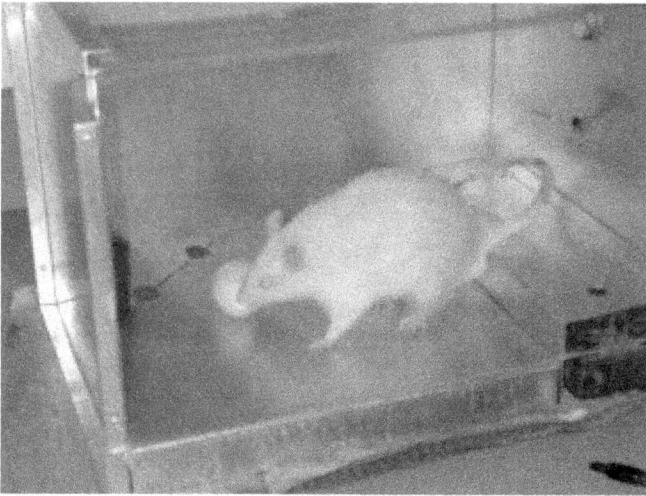

www.ingramcontent.com/pod-product-compliance
Lightning Source LLC
Chambersburg PA
CBHW051235200326
41519CB00025B/7381